DEEP THINKERS

DEEP THINKERS

Inside the minds of whales, dolphins, and porpoises

Edited by **JANET MANN**

CAMILLA BUTTI
HEIDI E. HARLEY
PATRICK HOF
VINCENT JANIK
ERIC PATTERSON
ANDREW READ
LUKE RENDELL
LAELA SAYIGH
HAL WHITEHEAD

The University of Chicago Press
Chicago

The University of Chicago Press, Chicago 60637

Published 2017

26 25 24 23 22 21 20 19 18 17   1 2 3 4 5

ISBN-13: 978-0-226-38747-5 (cloth)
ISBN-13: 978-0-226-38750-5 (e-book)
DOI: 10.7208/chicago/9780226387505.001.0001

Library of Congress Cataloging-in-Publication Data

Names: Mann, Janet, editor.
Title: Deep thinkers : inside the minds of whales, dolphins, and porpoises / edited by Janet Mann.
Description: Chicago : The University of Chicago Press, 2017. | Includes bibliographical references and index.
Identifiers: LCCN 2017013377 | ISBN 9780226387475 (cloth : alk. paper) | ISBN 9780226387505 (e-book)
Subjects: LCSH: Cetacea—Psychology. | Cetacea—Behavior. | Animal intelligence. | Cognition in animals. | Animal communication.
Classification: LCC QL737.C4 D398 2017 | DDC 599.515/13—dc23
LC record available at https://lccn.loc.gov/2017013377

This book was conceived, designed, and produced by
Ivy Press
58 West Street, Brighton BN1 2RA, United Kingdom

Publisher Susan Kelly
Creative Director Michael Whitehead
Editorial Director Tom Kitch
Art Directors Wayne Blades and Kevin Knight
Commissioning Editor Kate Shanahan
Senior Editor Stephanie Evans
Project Editor Fleur Jones
Designer Simon Goggin
Illustrators John Woodcock and Richard Peters
Picture Researcher Alison Stevens

Front cover image Bryant Austin

Printed in China

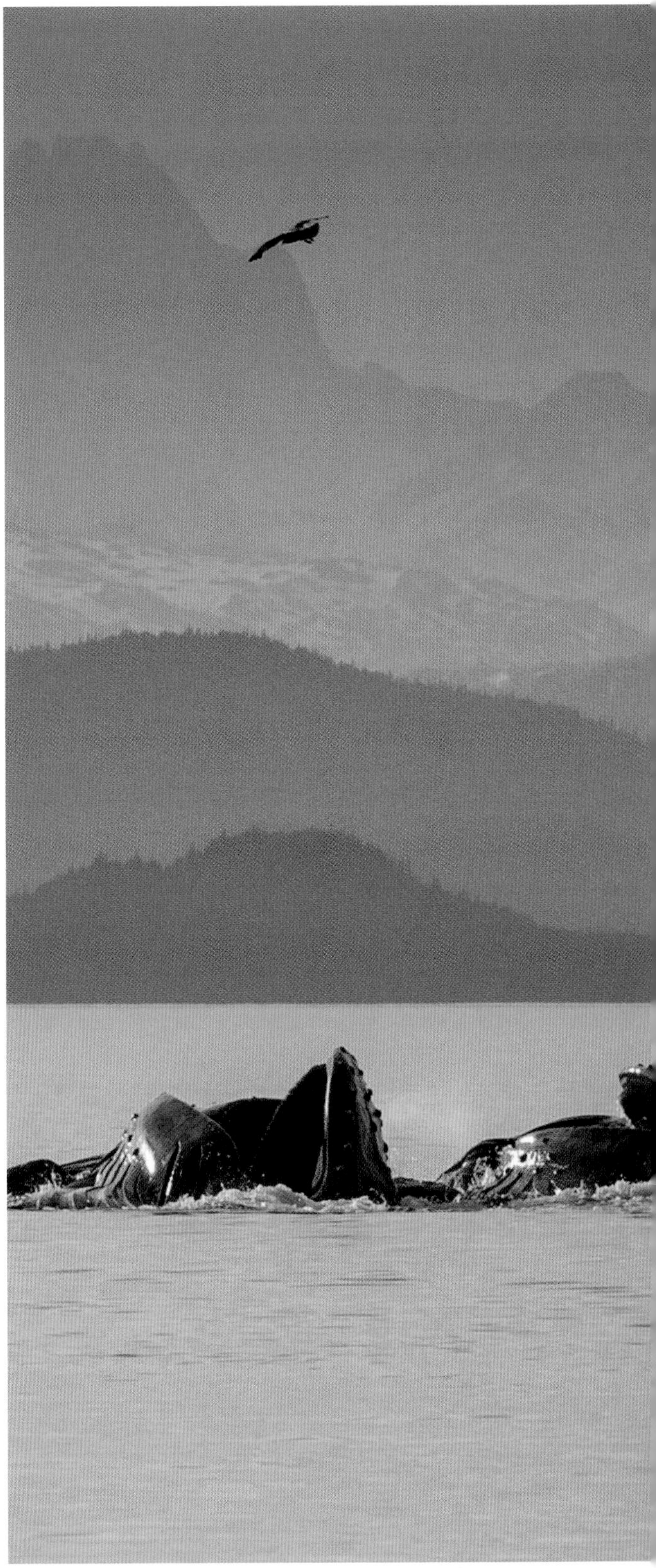

CONTENTS

INTRODUCTION Janet Mann

Few species delight humans more than cetaceans, with their massive size, acrobatic ability, social nature, and intelligence. Anyone who has been on a boat and seen a wild whale, dolphin, or porpoise remembers this experience for eternity, even if he or she knows little about them or even which species was spotted. Until quite recently, most of what we knew about cetaceans came from whaling, fisheries, or captivity. None of these circumstances were natural and all were focused on how cetaceans served humans (for food, oil, or entertainment), not on the animals themselves. However, the species we exploited so heavily for centuries now command our attention and respect for their extraordinary intelligence and evolutionary significance.

Since the 1960s, dolphins, whales, and porpoises have gained increasing legal protections from hunting, incidental takes (for example, netting a dolphin during fishing operations), and intentional captures for aquarium display. In 1972, the US Congress passed the Marine Mammal Protection Act, which established a moratorium on taking and importing marine mammals, albeit with some exceptions—such as aboriginal hunting. At the same time, wildlife biologists and comparative psychologists initiated studies of wild and captive cetaceans, with some long-term studies lasting more than 30 years. Scientists have contributed enormously to our understanding of these magnificent mammals. This work has revealed the intricacy and longevity of their social bonds, their elaborate networks and social structures, their resilient and ephemeral cultures, their complex foraging tactics, and their impressive cognitive skills. Importantly, this work also placed cetaceans in a comparative framework with other marine and terrestrial animals.

Our understanding of cetaceans has advanced considerably in the last 50 years and continues to change as we debate whether these highly intelligent animals should be in captivity at all, or should be afforded "rights" as sentient beings. Occasionally the popular press highlights an exciting research finding, but until now no book has covered the state-of-the-art scientific discoveries across cetacean species. To this end, some of the world's authorities on cetacean biology and behavior were invited to contribute to this book in their areas of expertise. Each chapter focuses on a key theme, concept, or problem. *Deep Thinkers* is dedicated to highlighting some of the extraordinary features of cetaceans that have entranced scientists and nonscientists alike.

WHAT ARE CETACEANS?

Modern-day cetaceans are impressive products of millions of years of evolution. Having undergone massive transitions since an ungulate-like (hooved) ancestor moved into the seas some 50 million years ago, cetaceans are an exemplary illustration of evolution at a grand scale and extreme adaptation. They also provide intriguing cases for convergent evolution—where a trait has evolved more than once to serve similar adaptive functions (like blubber in cetaceans and seals as both fat stores and insulation).

All cetaceans evolved critical adaptations to their marine environment. A marked increase in size let cetaceans reduce heat loss and maintain body temperature. The nose became a blowhole that migrated to the top of the head for easy breathing at the surface. For efficient movement in water, a number of physiological changes occurred including hair loss and replacing forelimbs with pectoral fins for steering and hind limbs with a robust tail and flukes for propulsion. The pelvis became detached from the vertebrae to facilitate movement. Cetacean blood and muscle is specially designed to retain oxygen through a high volume of blood, concentrated hemoglobin and myoglobin, and larger lungs, which can collapse when diving to avoid gas bubbles in the blood (decompression sickness or "the bends"). Cetacean eyes have anatomical and molecular adaptations to the light limitations of the aquatic environment, including a highly spherical lens and corneal design that allows for acute vision in both air and water. Their bones are dense—including the ear bones, which aid in underwater hearing. Cetaceans lost outer ears in favor of a streamlined physique, but they have fat pads in their jaws to aid in hearing. Because cetaceans nurse underwater, they can curl their tongues to enable nursing from inside the mammary slit.

HOW DO WE CLASSIFY CETACEANS?

About 39 million years ago, the ancestral cetaceans diverged into two groups reflecting vastly different physio-ecological, social, and life history strategies. These groups are the odontocetes (toothed whales, 75 species) and the mysticetes (baleen whales, 14 species). Odontocetes include a diverse range of species such as sperm whales, beaked whales, narwhals, killer whales, dolphins, and porpoises. Mysticetes include the largest animal on Earth, the blue whale, the crooning humpback whale, and smaller species such as minke whales. The physio-ecological, social, and life history traits that have evolved in cetaceans, and which are discussed below, also demonstrate convergence with other taxonomic groups in a range of other habitats (land, sea, air).

PHYSIO-ECOLOGICAL

As the name implies, odontocetes have teeth. These are mostly conical, but porpoises have spade-shaped teeth and the beaked whales and narwhals have tusks. Odontocetes mostly feed on individual fish or squid. The mysticetes have whalebone—known as baleen, which are keratinized plates specialized for filter feeding—and gulp large quantities of small prey, including fish and zooplankton. While odontocetes have a narrow rostra or beak and elongated jaws, mysticetes have a broad rostra and bowed mandibles to accommodate the baleen plates. Although some odontocetes, such as the sperm whale, are quite large, mysticetes are much larger, and adapted to making massive migrations from the polar regions to tropical waters, with distinct feeding and breeding cycles to correspond. Odontocetes sometimes have large ranges, but are not typically migratory per se. They range from the polar regions, where killer whales hunt marine mammal prey, to small rivers in Asia and South America. Many species of odontocetes have exquisite echolocation abilities to aid in prey capture.

SOCIAL

Most mysticetes are characterized as solitary (see Chapter 5), in that they rarely form stable associations with others. In contrast, odontocetes form some of the most stable groups known, from pilot whale and killer whale pods, to sperm whale matrilineal units. These are family units, although fathers are not known to associate with offspring with any regularity in any cetacean. Mysticetes are not known for establishing the long-term bonds that are relatively common in odontocetes. Thus, the communication systems of the two suborders are different, with mysticetes tending toward shared song across a population, a feature convergent with birds, and odontocetes using group-specific and individual-specific calls that are important for identification of social units and individuals (Chapter 4). Some odontocete species exhibit degrees of social complexity that are rare in the animal kingdom, such as long-term bonds, alliances, and fission-fusion (Chapter 5).

LIFE HISTORY

Although cetaceans are large-bodied to accommodate life in the sea, the mysticetes grow much faster and reach maturity earlier than odontocetes. The bowhead whale is an exception among mysticetes (discussed in Chapter 5). In addition, females are larger than males in most mysticetes, albeit slightly. One explanation is that larger body size is strongly favored for production of offspring. Among odontocetes, if one sex is larger, it is almost invariably males, suggesting that direct contest competition is important (Chapter 5). This is further evident with weaponry that only males have, such as the tusks of narwhals and the battle teeth of male beaked whales. Although there are 22 species of beaked whales, they tend to remain in open sea, and are the least understood of all cetaceans. Odontocetes grow slowly, sometimes only reaching maturity in their teens. While mysticetes nurse for less than a year, odontocetes nurse their young for more than a year. Similar to primates and other mammals with slow life histories, odontocetes

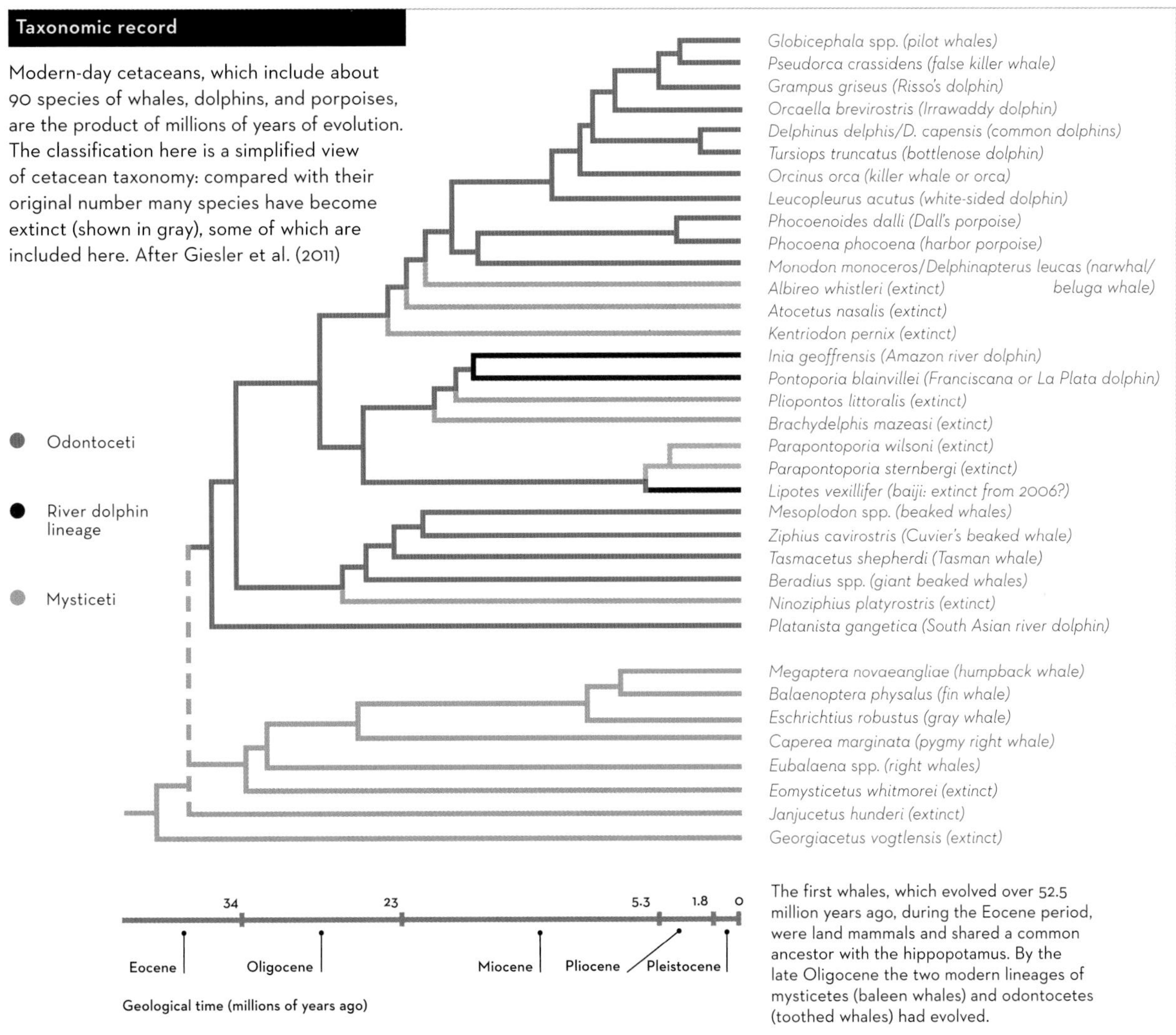

Taxonomic record

Modern-day cetaceans, which include about 90 species of whales, dolphins, and porpoises, are the product of millions of years of evolution. The classification here is a simplified view of cetacean taxonomy: compared with their original number many species have become extinct (shown in gray), some of which are included here. After Giesler et al. (2011)

The first whales, which evolved over 52.5 million years ago, during the Eocene period, were land mammals and shared a common ancestor with the hippopotamus. By the late Oligocene the two modern lineages of mysticetes (baleen whales) and odontocetes (toothed whales) had evolved.

have large brains—much larger than those of mysticetes—especially when controlling for body size (Chapter 2). The dolphin family (about 37 species) has the largest brains for their body size. The smaller porpoises (seven species), which are often mistaken to be dolphins, have relatively smaller brains and faster life histories. Evidence suggests there has been positive selection for increases in brain size in some taxa, such as the bottlenose dolphin.

CONVERGENCE

Remarkably, cetaceans have evolved features found in terrestrial and avian taxa to meet similar ecological or social circumstances. It is not just the physiological features that have evolved, but the molecular/genetic profile is similar between otherwise distant taxa. Genes important for echolocation and high frequency hearing in bats are also found in odontocetes. Genes associated with myoglobin concentration—essential for deep-diving ability—are similar in seals and cetaceans, despite very distant evolutionary histories. Within cetaceans, distinct families have converged on similar traits, such as three families of river dolphins occupying Asia and South America. Many other traits, such as increased brain size, unihemispheric sleep, song, blubber, and social behaviors have converged with primates, birds, and other mammals, but scientific investigation of functional genomes that correspond to these traits is just in the beginning stages. Evolution finds many ways to solve similar problems, but sometimes hits on the same solution over and over again.

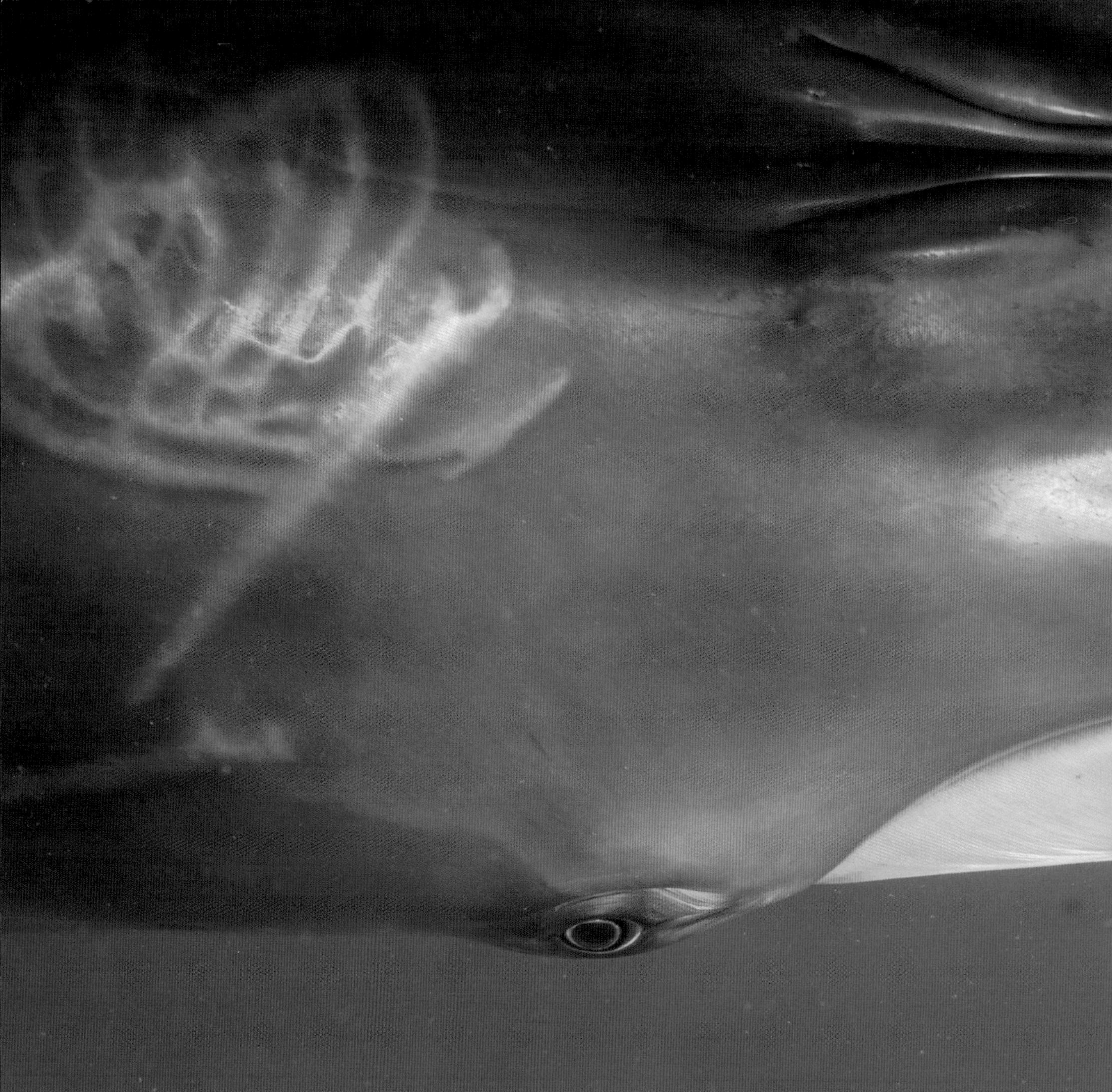

1 BENEATH THE SURFACE

Janet Mann & Andrew Read

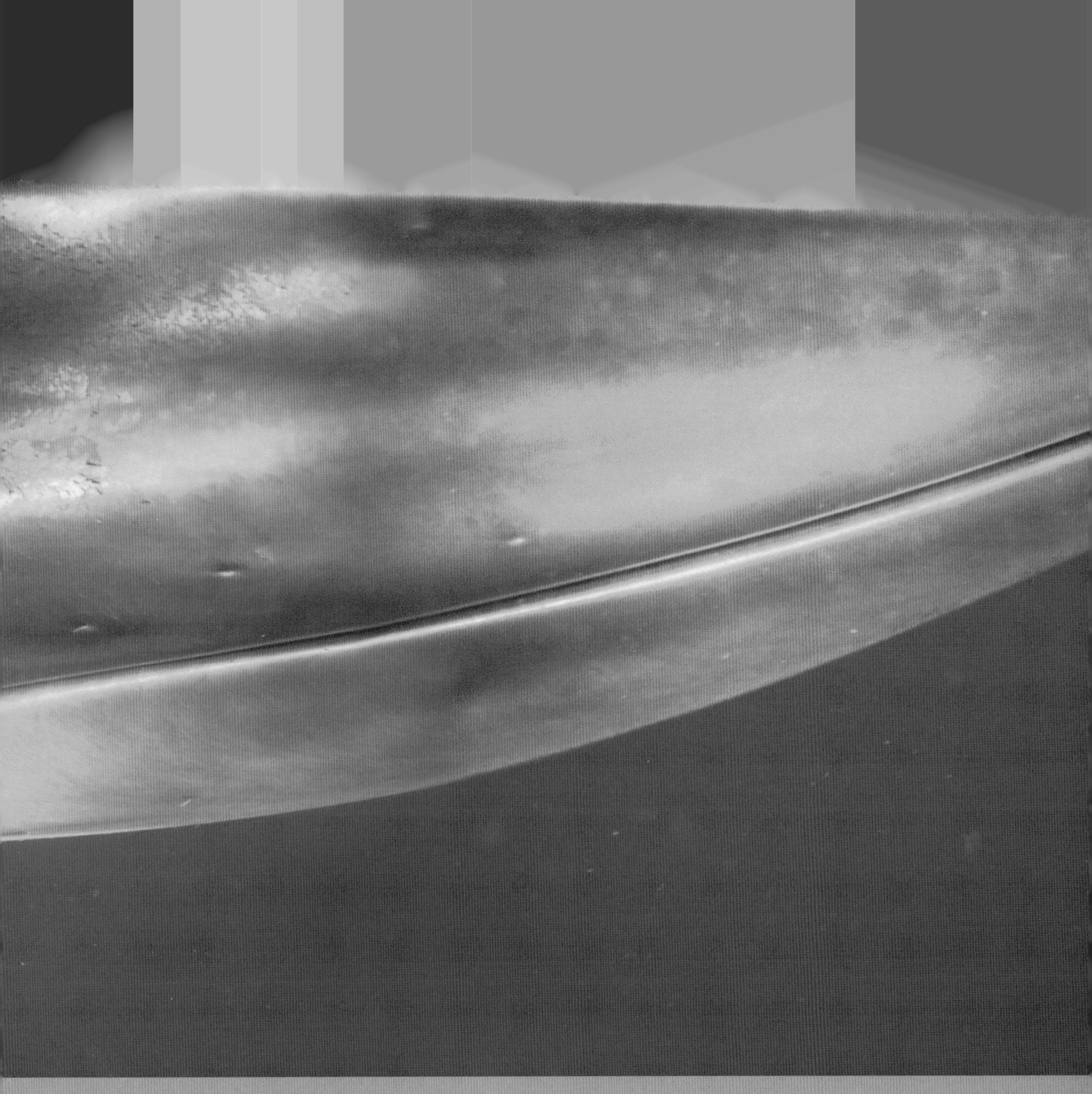

Dwarf Minke Whale Composite Portrait II
Great Barrier Reef, Australia, 2009

"This is a lifesize composite photograph measuring two by four meters. The final seamless print is comprised of five individual close-up photographs for which I used a digital medium format camera and portrait lens. The scarring on the whale's head is most likely the result of attack by a killer whale."—Bryant Austin

HOW & WHY DO WE STUDY CETACEANS?

Cetaceans are perhaps the most extreme mammals. In terms of social systems, life history, physiological, and behavioral traits, dolphins, porpoises, and whales represent a suite of unique adaptations to a fully aquatic lifestyle. It is these traits that inspire mythology and tales about how dolphins have rescued people from shark attacks, how dolphins can heal the sick, how sperm whales take their revenge on human hunters (as in the great story of the *Essex*, the story that inspired Herman Melville's *Moby-Dick*). But these traits also arouse the intense curiosity of scientists with the underlying question: how and why have these unusual creatures evolved such extreme traits?

When ancestral cetaceans moved into the seas fifty million years ago, they underwent massive and radical changes to accommodate their new marine habitat. The development of exquisite sensory systems, such as echolocation, allow odontocetes to navigate, hunt, and communicate in murky environments. Other senses, such as smell, virtually disappeared when their "noses" migrated to the top of their heads for breathing. Physiological traits, such as body size, organ size, limbs, and beaks, underwent colossal transformation.

The brain is a most impressive organ, and cetaceans have the largest brains on Earth (see Chapter 2). The sperm whale brain weighs nearly 18 lb (8 kg). But the sperm whale doesn't have the largest brain when body size is considered. That distinction, after humans, belongs to a large number of dolphin species, with rough-toothed dolphins at the top of the list, sporting a 3 lb 5 oz (1.5 kg) brain and an "encephalization quotient" (a measure of brain size scaled for body size) of 4.95 (about twice that of our closest relative, the chimpanzee, see Chapters 2 and 3). The biggest mammal ever to roam the Earth is the blue whale, weighing in at over 400,000 lb (about 200 tons, 180 tonnes). Cetaceans dive deeper and longer than any marine mammal, with some beaked whales reaching depths of 2 miles (3 km) and, remarkably, capable of holding their breath for more than two hours. And, cetaceans undertake the longest migrations of any mammal, with both gray and humpback whales sometimes traversing literally halfway around the world (see Chapter 6).

Another distinction of interest is testes size. Right whales have the world's largest testes, each about the size of a Volkswagen Beetle, weighing about 1 ton—a feature suggesting that males compete for fertilizations by sheer sperm count. Finally, the bowhead whale has the longest lifespan for any mammal, at more than 200 years. These extremes are fascinating in and of themselves, but they converge with other social and life history adaptations generally rare among mammals, but more common among highly social mammals such as primates, elephants, and cetaceans. Long-term study has revealed that cetaceans, and dolphins in particular, exhibit long-term social bonds, complex alliances, prolonged periods of maternal care, late maturity, and long lifespans. To understand such intriguing subjects, cetacean biologists have crafted innovative approaches to the study of animal social systems. This book explores why dolphins and whales have evolved such cognitive and social complexity and explains how we study these iconic animals.

Opposite Sperm whale off the coast of Dominica. Sperm whales are known for their extremely long head and narrow jaw. The head is filled with a waxy, fatty liquid, called spermaceti, which was coveted by whalers. Spermaceti is critical for transmitting sound during acoustic communication, particularly in the production of varied click-based codas that differentiate whale clans. Body size can also be determined by humans and likely other sperm whales, based on head size and intervals between pulses produced with each click.

CHALLENGES

When people learn that we study cetaceans, the first question is often "So, you swim with them?". When the answer is "no," the inevitable response is, "Then how can you study them?". Besides our lack of Olympian swimming and diving skills, there is simply no way a human swimmer can keep up with a cetacean engaged in normal activities for more than a few moments, let alone long enough to record their behavior. Cetaceans do not lend themselves to traditional observational methods typical of other studies of wild animals. They only spend a small proportion of their lives at the surface, so even the most dedicated biologist may catch only fleeting glimpses of their behavior. Cetaceans are also deep-diving and fast-moving, and sometimes migrate thousands of miles from feeding to breeding grounds; many species are only found far from shore. Following individuals for a day, let alone years, is a significant challenge. Residential coastal species, such as bottlenose dolphins, have been successfully studied in a few locations, but acquiring even basic information on most cetaceans requires creativity, tenacity, and patience. Here we illustrate how clever technologies and sheer persistence have allowed biologists to tackle the challenges of studying dolphins, porpoises, and whales.

WHALING HISTORY

Aristotle was fascinated by the biology of whales long before we knew their "economic" value, but the earliest studies of cetacean biology are firmly rooted in centuries of whaling. Fortunes were made on whale oil and baleen, so many of the keenest observers of dolphins and whales were the whalers themselves. Captains William Scoresby and Charles Melville Scammon wrote extensively about their voyages, detailed the behavior and life history of their quarry, and even speculated about their intelligence. Biologists working at whaling stations dissected massive carcasses and became familiar with their peculiar physiology and habits. One family of whalers, the Davidsons, was even successful in recruiting the assistance of killer whales (or perhaps it was the other way around) in Twofold Bay, Australia, where humans and killer whales worked together to kill migrating mysticetes (usually right whales). The killer whales ate the tongue and left the rest for the Davidsons. That relationship hinted at killer whale intelligence and cultural behavior more than a century before scientists had any inkling of these subjects (see Chapter 6). The Davidsons engaged in subsistence whaling and took fewer than a dozen whales each year, but commercial whaling was big business (see Chapter 8). By the

middle of the twentieth century, most populations of large whales had been decimated, fossil fuels replaced whale oil, and our attention turned to dolphins, which were maintained for the first time in captivity.

CAPTIVE RESEARCH

Although largely dedicated to entertainment, successes with breeding, maintaining, and training cetaceans in aquaria in the twentieth century provided opportunities for research as well, predominantly with three species: bottlenose dolphins and beluga and killer whales. In the 1960s, the Caldwells, a husband and wife team, discovered signature whistles in bottlenose dolphins (see Chapter 4), which sparked successful research on dolphin communication in captive and wild settings for decades. There was early promise in the research of John Lilly (see Chapter 3), who marveled at dolphin intelligence. Unfortunately, Lilly's work fell into disrepute when his experiments violated principles of ethical treatment with the use of psychoactive drugs and dishonored the scientific method itself. Few scientists dared to study dolphin intelligence or communication after Lilly's speculative, but wildly popular publications. But one scientist stood out and almost single-handedly redeemed studies of dolphin cognition, Dr. Louis M. Herman, an experimental comparative psychologist who conducted some of the most original and important experimental work on dolphin cognition (see Chapter 3). His methods were based on behaviorist traditions, using operant conditioning—such as positive and negative reinforcement, so that dolphins learned to associate a category of responses with rewards (fish). But his discoveries went well beyond traditional learning paradigms. Herman's research institute, the Kewalo Basin Marine Mammal Laboratory (KBMML) in Honolulu, Hawaii, differed from most other cognition laboratories in that it was solely dedicated to research. Today, several aquaria allow for research, but the animals are also for public display, sometimes "swim-with" programs, and entertainment.

Opposite Nineteeth-century engraving of a whale hunt. The whale depicted here is likely to be a right whale, so named by whalers who saw them as the "right" whale to hunt, being easily approached, slow swimming, found close to shore, and floating when dead to ease the task of accessing their valuable oil, meat, and whalebone (baleen). As a result, the species became perilously close to extinction.

Below Lou Herman working with dolphins Phoenix and Akeakamai. Lou is holding an underwater speaker through which imperative sentences were played to Phoenix and arbitrary sounds for Akeakamai to imitate. From 1970 to 2004, Lou, together with his students and colleagues, conducted groundbreaking research at KBMML on many aspects of dolphin cognition including studies of visual acuity, acoustic discrimination, echolocation perception, short-term memory, acoustic and motor imitation, concept formation, memory for self-performed actions, body-part awareness, understanding human-directed pointing and gazing, understanding televized images, and artificial language comprehension.

FIELD RESEARCH

Exciting new field research methods have provided a window into dolphin and whale societies that was unimaginable only a few decades ago. Since the 1970s, scientists have moved from making inferences from dead whales killed for profit to the long-term study of individuals over multiple generations. This exciting work has opened new insights into the complex social world in which these animals live, in which the nature of social relationships can determine success or failure of an individual over the course of its lifetime.

INDIVIDUAL IDENTIFICATION

The study of wild cetaceans over the long term is a fairly recent phenomenon. Two long-term studies were established in the 1970s and both continue to this day. Randy Wells pioneered research on bottlenose dolphins in Sarasota Bay, Florida, and Michael Bigg and his colleagues began to study killer whales in Pacific waters off British Columbia. The key to these long-term studies was the ability to identify individual animals through photographs, a method often described as photo-ID. In 1977, Bernd and Melany Würsig published a landmark paper in the journal *Science* showing that the dorsal fins of bottlenose dolphins in Argentina were individually distinctive and could be used to follow individuals across years. Biologists quickly found numerous ways to identify their subjects, from patterns of saddle patches and dorsal fins on killer whales, to fluke pigmentation on humpback whales, to right whale callosity patterns. By using these natural markings, researchers could identify their study subjects with nothing more than a good-quality photograph.

SURVEYS

Observational methods were quickly developed to determine which animal was associating with whom and where. These survey methods can be opportunistic, recording the identities of dolphins or whales where and when you find them, or can use more systematic transect methods, in which observers move along a predetermined line and sample all groups within some distance from that line. Such data are useful for determining the distribution, abundance, and habitat use of individuals and populations. Most surveys are boat-based, but some studies of migrating whales or of cetaceans in small bays are collected from vantage points on shore. Aerial surveys (and more recently drone-based imagery) can be used to study dolphin and whale populations over larger distances. During a survey, observers typically record the location (latitude and longitude) of a group, photograph all individuals, and determine their general behavior, which typically falls into four types: foraging (feeding), resting, socializing, and traveling. Group membership is usually defined by distances between individuals (for example, all whales within a hundred yards of each other are part of the same group). Ecological data might be collected such as water depth, habitat, and sea state. As will be discussed extensively in Chapter 5, surveys are extremely useful for determining social structure and the strength of bonds within a population.

FOCAL SAMPLING

In the 1980s, scientists initiated a study of wild bottlenose dolphins in Shark Bay, Western Australia, a remote and unspoiled UNESCO World Heritage Site. Here, in shallow and pristine waters, scientists could observe the fine-scale details of dolphin behavior and social interactions, much like other scientists had achieved from the study of terrestrial species such as primates, lions, elephants, and hyenas.

The Shark Bay dolphins have fairly restricted ranges, so it was relatively easy for scientists to stay with an individual dolphin for many hours using a small boat. This allowed researchers to use a method known as "focal sampling" in which the behavior of an individual or a mother-calf pair is systematically collected over a study period. For example, an observer might record the behavior of the mother, her calf, and the distance between them every minute for a period of several hours. Bottlenose dolphins live in a dynamic "fission-fusion" society (see Chapter 5), meaning that membership of their group changes

throughout the day. Focal sampling enabled observers to track these changes from an individual's perspective, while collecting detailed information on location, habitat, and other important contextual details. The same individual can be followed on many days, allowing researchers to build a more complete picture of their behavior and enabling us to address questions such as: Are there individual differences in maternal care? Does maternal behavior change based on experience or the sex of the offspring? Focal sampling is also useful for examining the detailed nature of social interactions, such as who plays, fights with, or grooms each other.

Above Focal sampling is the technique used by researchers to identify and catalog individual cetaceans. Humpbacks (top left) have distinctive patterns of black and white pigmentation on their tail flukes as well as unique scarring and shading patterns. Southern right whales (top right) develop huge growths or callosities on their heads and rostrum (the upper jaw). The pale color is derived from the presence of whale lice (a type of crustacean) that live on the callosities. Fins are other obvious markers: the badly damaged dorsal fin identifies a bottlenose dolphin studied—in Shark Bay, Australia (bottom left), and (bottom right) killer whales have individually distinctive white eye and saddle patches near the dorsal fin in addition to other markings.

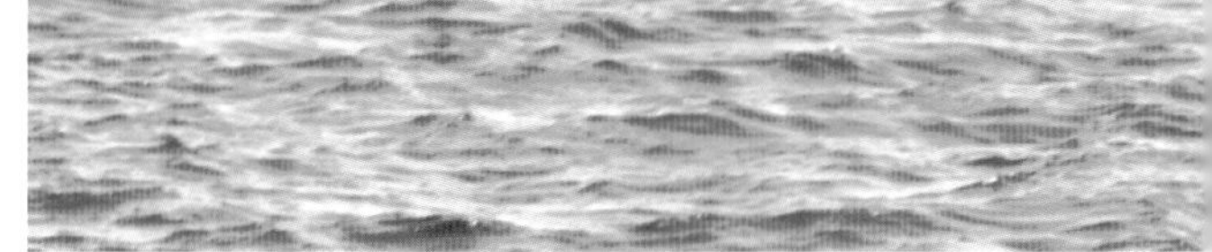

CASE STUDY: **FOCAL SAMPLING**

In the late 1980s, Janet Mann began a long-term study of bottlenose dolphin adult females and their offspring in Shark Bay, Western Australia. With a background in primates, and training from Jeanne Altmann who pioneered standardized quantitative behavioral methods, including focal animal sampling, Mann began following mother–calf pairs for 1–9 hours a day from a 13-ft (4-m) dinghy. She began this work with her graduate school advisor, Barbara Smuts, also a primatologist.

DATA COLLECTION

Data collection involved a list of mothers and their nursing calves, which were searched for and subsequently observed for multiple days each year from birth through the period of weaning. Systematic data were collected every minute, including mother-calf proximity, activity, and what other dolphins were present. Location and habitat were recorded as well. Detailed events, such as petting (flipper stroking of another dolphin), synchronous breathing, and other interactions were recorded for both mother and calf.

MOTHER–CALF PAIRS

Early on, two mother–calf pairs stood out: Crooked Fin and her son Cookie, and Yogi and her son Smokey. Both mothers were loners and spent little time with other females, but Cookie and Smokey would frequently separate from their mothers, traveling hundreds of yards away to join up with each other and occasionally other young males, also separate from their mothers, particularly Urchin, another calf in the study. They would play very intensively and practice synchronized displays, even in the first year of life. Socio-sexual behavior such as mounting, goosing (beak to genital), and petting and rubbing were common, sometimes with or without a young female partner present. These interactions were reminiscent of adult male alliance behavior, where males (usually 2 or 3 but up to 14) cooperate against other alliances to gain access to cycling females. While such behavior might be important for adult males, it was striking to see this behavior in calves so young, more than a decade away from establishing such relationships. Mann decided, when mothers and calves separated, to remain with the calf and monitor the mother from afar, who was invariably foraging alone. Group composition was recorded every minute using a 33-ft (10-m) chain rule (dolphins within this distance of another group member were part of the group), and who joined and left the group was noted. The shallow clear water allows observers to see what the dolphins are doing. Cookie and Smokey were observed on multiple days each year.

COOKIE AS CALF, 0–3.5 YEARS

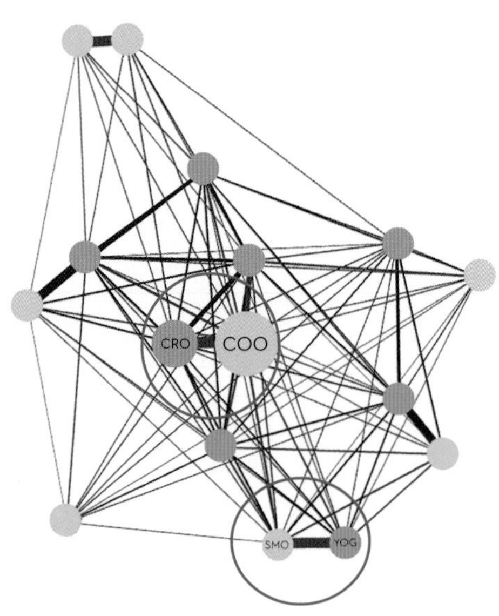

COOKIE & SMOKEY ALONE

In 1992, Cookie and Smokey were suddenly orphaned—their mothers died within a month of each other. As both calves were now over three years old (typical weaning age), they survived their mothers' deaths. Strikingly, they now spent almost all of their time together, in very close contact, such as touching flippers ("holding hands") or

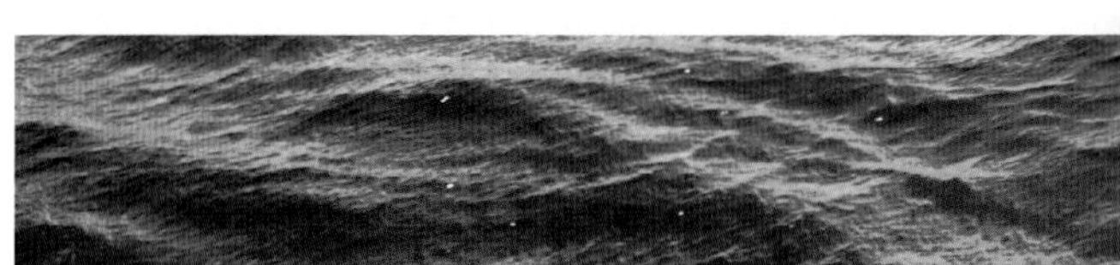

THE ORPHANED COOKIE AS YOUNG JUVENILE, 3.5–7 YEARS

AS OLDER JUVENILE, 8–11 YEARS

AS YOUNG ADULT, 12–20 YEARS

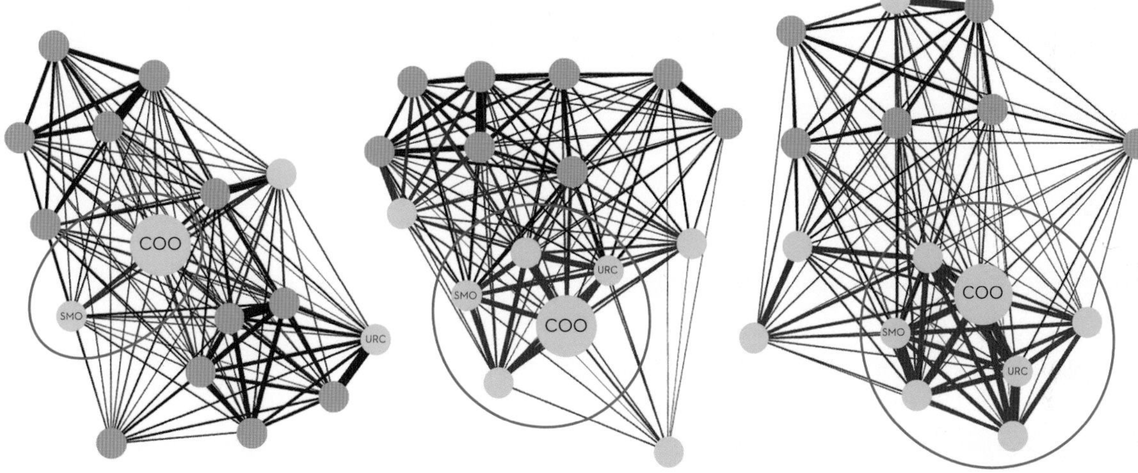

resting a flipper on the other's side. We now know that the period right after weaning is risky for young males who are often the object of aggression from juvenile and adult males. By staying together, the orphans likely ensured their mutual survival. Mann continued conducting focal follows on both of them for several more years and continued to track them through surveys thereafter. Although Cookie and Smokey had ups and downs in their relationship, today—30 years on—Cookie, Smokey, and Urchin are in a very tight adult male alliance, applying the skills they began to develop decades ago. Their remarkable story exemplifies the intricacies, investment, and importance of long-term social bonds, and what such bonds can mean for survival and reproduction.

- Females (CRO = Crooked Fin, YOG = Yogi)
- Males (COO = Cookie, SMO = Smokey, URC = Urchin)
- Line width indicates strength of ties between male and female dolphins in the group

Opposite and above The ties and interactions of Cookie, son of Crooked Fin, from birth to adulthood. As a calf (1) Cookie's top associates include three mother-son pairs. Smokey was his strongest partner. His mother died around the time he was weaned (2), and his bond with Smokey grew stronger. Cookie was with female cliques often, including his sister and niece. By the late juvenile period (3), strong bonds with males are evident, although still connected to females. By early adulthood (4), Cookie has strengthened his ties with males even more. Females become more peripheral to his network.

BIOLOGICAL SAMPLING

Other methods of biological sampling have become important to determine the health, relatedness, paternity, and genetic structure of populations. Researchers now commonly employ remote sampling methods, by using a crossbow or air rifle and a surgical stainless-steel biopsy tip to take tiny samples of skin to look not only at genes but at which genes are being expressed. Biopsy samples can also be used to assess hormone levels in blubber (akin to a human pregnancy test) and to age individuals (for example, telomere length, DNA methylation patterns in skin, and aspartic acid racemization). And the samples provide insight into what the sampled animal has been eating and into the quality of its blubber, which reflects its overall physical condition. We can estimate the age of some individuals from color patterns. For example, some dolphins, such as Indian Ocean bottlenose dolphins and spotted dolphins, speckle with age, so the degree of speckling indicates relative age. Other noninvasive biological sampling methods include "blow sampling," where scientists capture the "blow" or respiratory vapor from dolphins or whales using a long pole or even specially equipped drones, also known as unmanned aerial systems. This information can be used to look at the microbiome (similar to how we research microbes in the gut), dolphin DNA, and hormones. And we collect poop, to look at what the whales have been eating and to examine stress hormones. Specially trained dogs are sometimes used to detect floating whale feces, picking up their pungent scent. No stone is left unturned, as the saying goes.

ACOUSTIC SAMPLING

Dolphins and whales are among the most vocal of all mammals and we can listen to the sounds they produce underwater using a variety of methods, commonly referred to as passive acoustic monitoring. As noted above, many dolphins produce signature whistles, allowing us to identify individuals based on the frequency-modulated tones they produce. Because sound is so important to these animals, researchers have developed digital acoustic tags, known as DTags, that can be attached to study subjects with suction cups for periods of up to a day. These tags function much like your smartphone, with the capacity to record sound and measure acceleration, orientation, and depth. DTags provide us with the opportunity to follow and listen to our study animals for brief periods. They have proved

Below left "Blow sampling," where scientists collect dolphin respiratory vapor with a funnel attached to a test tube. Water samples are collected at the same time to act as a control since blow samples inevitably contain water. The wonderful thing about blow sampling is that it can be done repeatedly for the same individual on different days.

Below Drones can be used to collect blow samples and also to conduct surveys and collect behavioral and health data. Whales (here, humpbacks) and dolphins do not react much to drones hovering a few feet above.

invaluable in providing windows into the lives of animals, such as beaked whales, that are difficult to study using traditional methods. Concern over the potential effects of various types of ocean noise, such as military sonar, have led researchers to conduct playback experiments, in which animals wearing DTags are exposed to certain types of sound (see Chapter 8). Such experiments can also be conducted with natural sounds, such as the calls of mammal-eating killer whales, to determine how individual dolphins and whales respond to the calls of predators and other animals. Researchers also employ sophisticated bottom-mounted acoustic recorders to record the sounds of whales and dolphins in remote environments, such as the Antarctic, where visual surveys are difficult or impossible to conduct.

NEW TECHNOLOGIES

In recent years, a plethora of technological advances have enabled us to tag, track, biologically sample, and observe wild cetaceans in ways that reveal their hunting behavior, diving ability, patterns of movement, and social interactions and structure. Importantly, these methods can also measure how cetaceans respond to human disturbance. Drones allow us to survey whales and dolphins over vast areas through aerial videography. On a smaller scale, drones can be fitted with devices to capture blow. In addition, drone cameras can be used to examine behavior and the relative size, fatness or condition of whales and dolphins. In some cases, drones can be used to identify individuals. Tags are used to measure real-time physiological patterns (heart rate and swimming speed), just like human "fitness trackers." Satellite-linked tags reveal that humpback whales can travel more than 10,000 miles (16,000 km) across an ocean basin and that beaked whales dive to a depth of nearly 2 miles (3 km). The most recent advanced tags now record sound, acceleration, and orientation. Deep-diving pelagic cetaceans remain poorly studied and we will continue to rely on advanced technology to study their behavior. In the lab, digital software is improving our ability to identify and match photographs of individuals within and across geographic areas; computational power has enabled us to analyze the complexity of large dolphin communities using social network analysis. Social behavior of wild cetaceans remains a final frontier because we typically tag one animal at a time and see only its point of view. Long-term studies are integrating these advances together with the painstaking traditional methods of observation. And, despite all of these wonderful advances, we have merely scratched the surface of the rich lives of these fascinating animals.

Below left A beluga whale, tagged with a satellite transmitter, off the Northwest Territories, Canada.

Below Researchers radio-tag an Atlantic white-sided dolphin before releasing it to gather data on its daily movement and dive behavior.

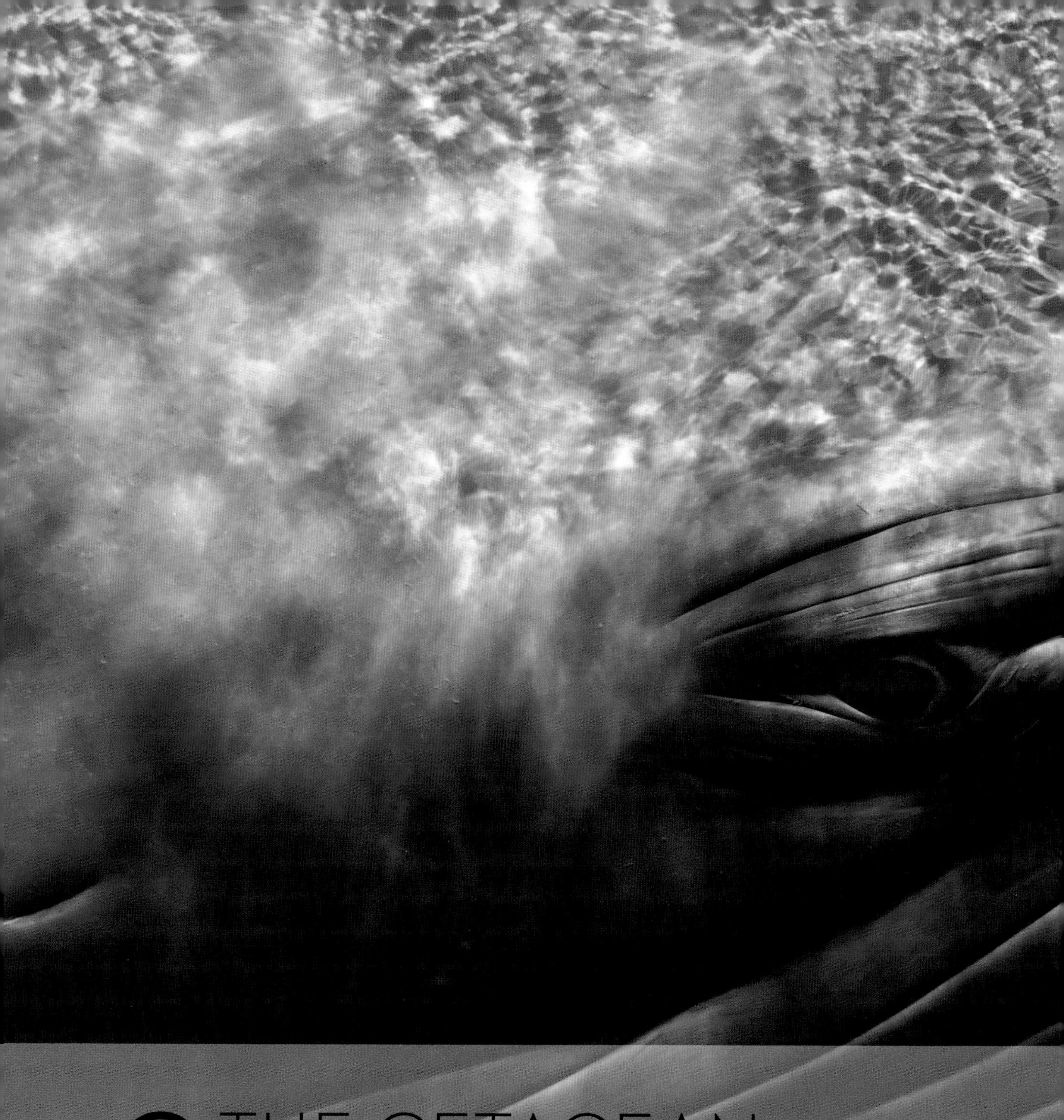

2 THE CETACEAN BRAIN

Camilla Butti & Patrick Hof

Dwarf Minke Whale Composite Portrait I
Great Barrier Reef, Australia, 2009

"I spent over twenty hours with this whale, allowing me to create my largest and most detailed photographs of whales. This whale was given the name Ella by researchers who first discovered her in 2006."—Bryant Austin

EVOLUTION OF THE CETACEAN BRAIN

The brain is the control center of the body, collecting sensory information and transforming it into the behavioral responses necessary for survival. Changes in the environment and biology of a species are reflected in changes to the shape, size, and structure of the brain. This is demonstrated by the tremendous variation in brain size and structure among vertebrates, including the degree of cortical folding, and the size of the visual, motor, somatosensory, and auditory fields.

ADAPTING TO LIFE UNDERWATER

During the evolutionary transition from land to water, more than 55 million years ago, the ancestors of cetaceans had to face several major challenges while adapting to a fully aquatic lifestyle: how to breathe, how to maintain a constant body temperature, how to improve the hydrodynamics of a four-legged body, and how to communicate and hear underwater. To overcome these problems, several bodily modifications took place including the regression of hind limbs and the pelvic arch, the development of a thick layer of blubber to maintain body temperature, the remodeling of the skull to migrate the nasal openings to the top of the head to allow for breathing in an aquatic environment, and the remodeling of the upper respiratory tract to act as a sound transmitter.

CHANGES IN THE BRAIN

Along with major transformations in body size and shape, a less obvious but extremely significant set of transformations occurred within the brain, including overall size, size of specific brain regions, external shape, and structure. While the majority of the modifications to the body were adaptations to an aquatic environment, the adaptations to the size and structure of the brain have been the object of speculation about cetacean intellectual abilities, based on evidence for tool use, mirror self-recognition, abstract rule comprehension, memory, and vocal and behavioral mimicry (see chapters 3 and 7).

Recent studies have shown that during evolution the mass of cetacean brains increased relative to body mass, known as encephalization (measured by the encephalization quotient, EQ), followed by an increase in complexity and the degree of neocortical convolution. The brain of the hippopotamus, the closest living relative of dolphins and whales, differs considerably from that of cetaceans, suggesting that the evolution of the cetacean brain was influenced by the specific selective pressures of adapting to a fully aquatic lifestyle.

In odontocetes, encephalization increased significantly and modern odontocete species possess values of EQ that are comparable to those of nonhuman primates and second only to humans. Modern mysticetes have brains that are much smaller, relative to their body size, than those of odontocetes. In mysticetes, the increase in encephalization is obscured by a disproportionate increase in body mass that was not accompanied by an increase in brain size. So while energetically expensive brains show positive selection in odontocetes (that is, favored by natural selection), constraints on body size were relaxed for mysticetes with no concomitant selection pressure on brain size. This means that while in odontocetes brain and body mass did increase simultaneously, the rapid increase in body mass seen through evolution in mysticetes was not followed by a simultaneous increase in brain mass. Consequently, mysticetes have smaller brains than expected given their body size.

Comparing hippopotamus & cetacean brains

Hippopotamus and cetacean brains differ extremely in external morphology. However, similarities in the neuronal organization and morphology support the phylogenetic relationship between hippopotamuses and cetaceans.

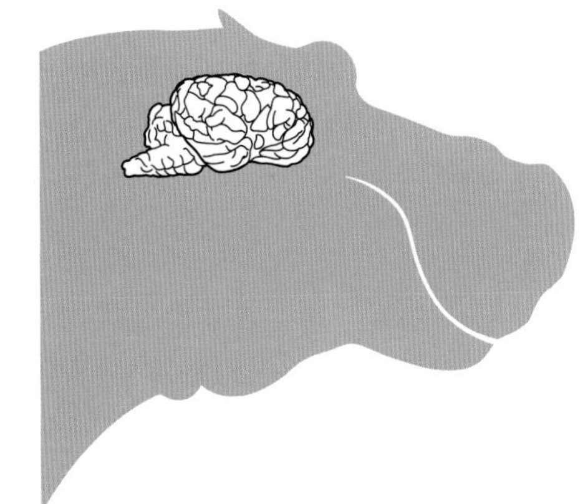

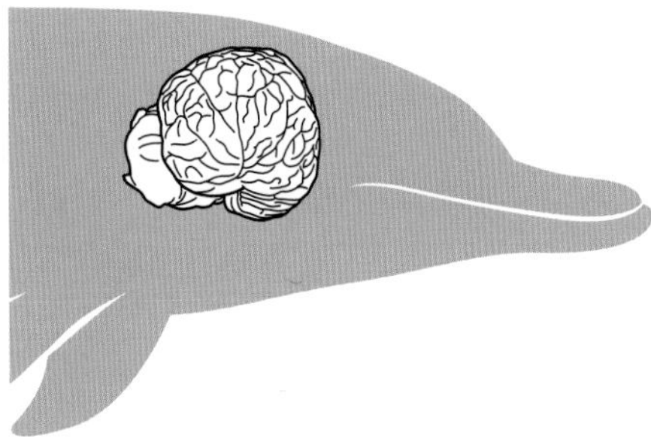

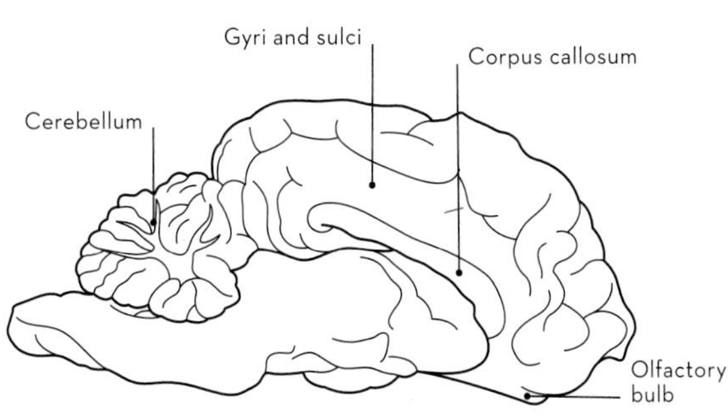

HIPPOPOTAMUS MIDLINE

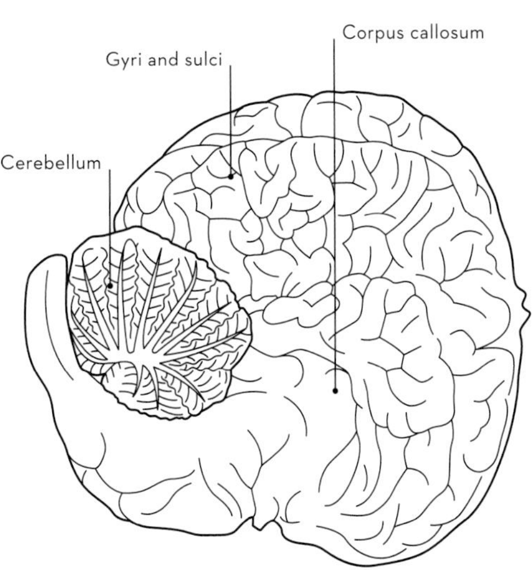

DOLPHIN MIDLINE

GYRI AND SULCI
Large brain size and a high degree of convolution are unique to cetaceans and not shared with hippopotamuses. These ridges and furrows in the cortex are known as gyri and sulci respectively.

CORPUS CALLOSUM
This structure forms the major link between the two halves of the brain and is larger in hippopotamuses relative to total brain mass. The cetacean corpora callosa have shrunk in size during evolution.

HIPPOCAMPUS
This structure is relatively larger in hippopotamuses than in cetaceans and these differences may relate to spatial memory and underwater navigation. As cetaceans appear to rely less on spatial cues than terrestrial mammals, they probably use different means for navigation (e.g., echolocation, brain magnetite for geomagnetic guidance during migration).

OLFACTORY BULB
This structure, present in hippopotamuses, is not evident in odontocetes as structures associated with processing sense of smell have shrunk during evolution in the latter. Mysticete species possess only a rudimentary olfactory bulb.

CEREBELLUM
The cerebellum is much smaller in the hippopotamus brain than that of cetaceans. The large size of the cetacean cerebellum is linked to the size of the neocortex.

ANATOMY OF THE CETACEAN BRAIN

We humans pride ourselves on our large brains and, with the unique ability to express our thoughts, feelings, and memories in a structured language, consider ourselves to be the smartest creatures on Earth. Based on our current definitions of intelligence, we eclipse orangutans, gorillas, and chimpanzees, and many other large-brained mammals such as cetaceans and elephants. However, we still have a fascination with understanding their special capabilities and comparing them to our own.

What makes whales, dolphins, and porpoises so special in our eyes? We might be tempted to say that cetaceans are smart because they have large brains. In part this is true; in fact, the largest brain on Earth belongs to an odontocete, the sperm whale, and weighs about 17 lb 9 oz or 8 kg). However, the human brain only weighs about 2 lb 12 oz (1.25 kg), so while it may appear that size matters, it cannot be everything. So what confers intelligence besides absolute and relative brain size? This is where the cellular structure and shape of the brain come into play. Because brain tissue is energetically expensive to maintain, brain reorganization may be an efficient strategy to increase intelligence rather than just increase brain size per se. Recent studies have indeed demonstrated that the cetacean brain has a complex cellular structure.

CORTICAL FOLDING

The cetacean brain is extremely folded and large, both in absolute terms and relative to body size, and is rotated toward the beak and toward the belly. The extreme folding of the cortex results in a neocortical surface larger than that of any other mammal, including humans. The cetacean neocortex is composed of a series of ridges and furrows known as gyri and sulci, which are comparable to those of hoofed mammals (related by common ancestry) and are organized around a vertical fissure known as the Sylvian fissure. The gyri and sulci are used as landmarks to identify the motor, visual, auditory, and somatosensory areas in the brain. Scientists have mapped these major functional areas, but there is still a significant portion of the neocortex that is unmapped.

THE DIFFERENCES BETWEEN PRIMATE & CETACEAN BRAINS

Compared to land-based mammals the cetacean brain has extensive parietal, temporal, and frontal cortical regions. The frontal part of the cortex, known as the frontopolar cortex, is extensive and may correspond to the prefrontal cortex of primates, which is a region involved in social behavior, decision-making, and the expression of personality. On the other hand, the hippocampus, which is involved in learning and memory, is extremely reduced, suggesting that in dolphins and whales the mechanisms of learning and memory are very different from those in other mammals. It may be related to the fact that the food sources they rely on are highly mobile, and as such, in contrast to terrestrial mammals, cetaceans do not need to remember where food patches are available in a given habitat using spatial cues, but they have to remember social affiliations that are important.

The diminutive hippocampus of cetaceans is believed to be an adaptation to their fully aquatic lifestyle. Differences in the development of the hippocampus in cetaceans and terrestrial mammals could be related specifically to differences in spatial memory and navigation in the water and they could reflect migratory patterns, as seen in birds. It has also been speculated that some cetaceans, and particularly mysticetes, rely on an ability to detect and analyze the direction of the Earth's magnetic field for navigation, although there is little direct evidence for such use of a geocompass. Contrasting features of the hippocampal formation, such as a very well-developed entorhinal cortex and its diminutive projection area, the dentate gyrus, and prominent cortical regions in cetaceans (for

Comparing the brains of dolphins & humans

Dolphins possess large brains and they have several things in common with large brain species such as humans and great apes: they live long lives, form stable communities, and show total parental dependence during childhood.

DOLPHIN DORSAL, LATERAL & MIDLINE

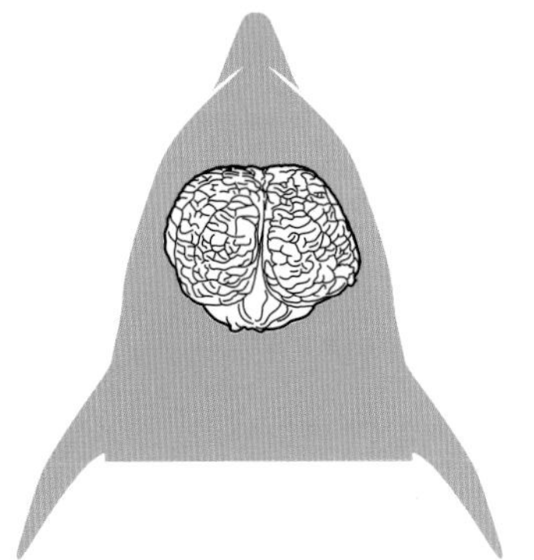

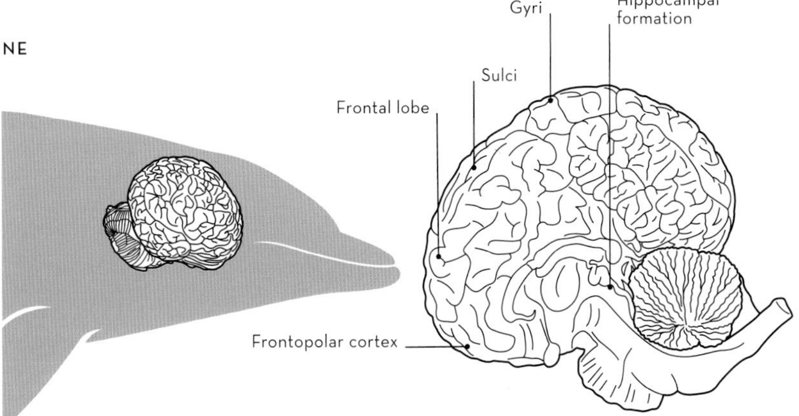

SULCI
The cetacean brain has a complex pattern of folding, or gyrification, and includes a series of gyri (ridges) and sulci (furrows) that create a brain structure considerably different from that of primates.

FRONTAL LOBE
The frontal lobe is rotated rostro-ventrally in cetaceans and includes regions involved in cognitive, emotional, motor, and executive functions.

PREFRONTAL (FRONTOPOLAR) CORTEX
Even though a region that corresponds to the human prefrontal cortex has not yet been identified in cetaceans, this is likely to be the frontopolar cortex, a large and highly folded area located in the rostro-ventral portion of the brain.

HIPPOCAMPAL FORMATION
The hippocampal formation presents contrasting features in cetaceans. Their small hippocampus and large entorhinal cortex have led scientists to think that the networks underlying memory processing in cetaceans might work very differently to those in other mammals.

HUMAN DORSAL, LATERAL & MIDLINE

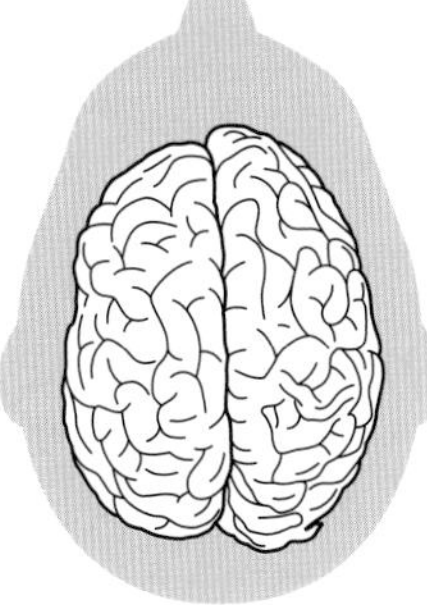

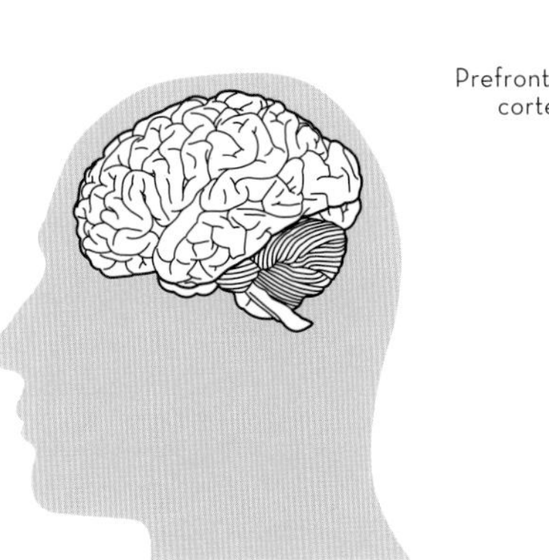

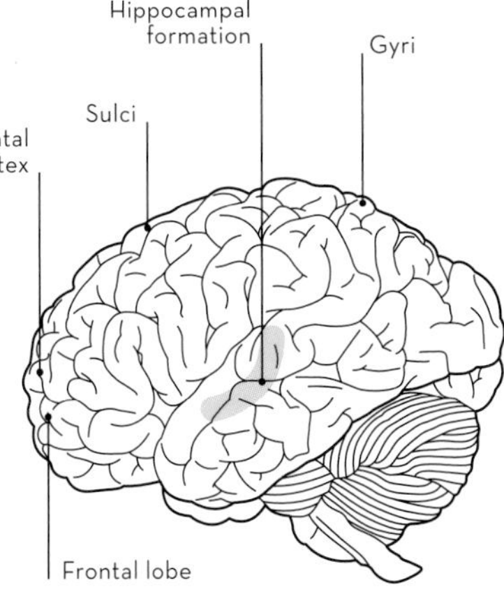

SULCI
The human brain is highly convoluted. The complex folding of the cortex increases the surface area available.

FRONTAL LOBE
The frontal lobe includes the frontal sections of each half of the brain and includes regions involved in planning, motivation, cognition, and executive functions.

PREFRONTAL (FRONTOPOLAR) CORTEX
The prefrontal cortex is involved in cognitive and memory-related processes, attention, and executive function such as judgment and the planning of complex behaviors.

HIPPOCAMPAL FORMATION
The hippocampal formation includes the hippocampus, the entorhinal cortex, and the dentate gyrus (none of which can be identified from the midline), and is involved in short-term memory, long-term memory, and spatial navigation.

Above The dominance of the sense of hearing in dolphins is striking. Their brains need to process a large amount of acoustic information while accounting for the high speed of propagation of sound in water. Although the use of vision is limited, acoustic and visual inputs are integrated within the neocortex during echolocation (scanning of the environment with a sonar system), and purely acoustic inputs are processed during communication between individuals, particularly important when hunting in groups in the hundreds or even thousands.

example, cingulate and insular cortices) may play a role in the mechanisms of learning and memory in these species. In other words, other cortical regions might have adopted some of the typical functions of the hippocampus in cetaceans.

INTERHEMISPHERIC CONNECTIVITY & AUDITORY STRUCTURES

The corpus callosum, which forms the major link between the two halves of the brain, is smaller in cetaceans than in other mammals, relative to overall brain mass. This may be due to the functional independence of cetacean brain hemispheres, which allows for each hemisphere to sleep independently—known as unihemispheric sleep, an unusual form of mammalian sleep characterized by the absence of rapid eye movement (REM) sleep, and the ability to swim while sleeping and with only one eye closed at a time. Unihemispheric sleep seems to be linked to the necessity of surfacing to breathe and maintaining vigilance during sleep. REM or "dream" sleep is incompatible with paying attention to one's external surroundings. In odontocetes the corpus callosum is smaller in large-brained species and its size is not compensated for in other connective structures. Connectivity between brain hemispheres reduces as brain size increases because larger brains have a lower density of neurons and it seems that fewer nerve fibers are needed to establish interhemispheric connections.

Brain structures that relate to the processing of acoustic information are particularly well developed in cetaceans, while structures that relate to sense of smell are completely lost in odontocetes and reduced to rudiments in mysticetes. The cetacean thalamus is extensive and contains large areas that are involved in the auditory system, reflecting the importance of hearing and communication for cetacean species.

The cetacean cerebellum represents about 15–20 percent of total brain mass. The large cerebellum size may be linked to the extreme expansion of the neocortex and the need for complex acoustic-motor processing. This is evident to anyone who has observed a dolphin echolocating on and chasing a fast-moving evasive fish or engaged in elaborate play and synchronous behavior.

CORTICAL LAYERS

The primate cortex is made up of six layers (I–VI, shown on page 32); however, the cetacean cortex is composed of only five layers and lacks a layer IV. This is puzzling because layer IV is important for communication between the thalamus (responsible for many regulatory functions, including sleep and consciousness, and sensory-motor signals) and the cortex. The cortex in both odontocetes and mysticetes includes a thick layer I that represents about one-third of the total cortical thickness, a thin and densely packed layer II containing neurons that extend into the first layer, a wide layer III containing small pyramidal neurons, a layer V containing large pyramidal neurons, and a layer VI that includes neurons of different shapes and sizes. The absence of layer IV in the cortex of cetaceans might be compensated for by a specific wiring pattern of the cetacean cortex characterized by a shift of thalamic inputs to the thick layer I, rather than deep layers (layer IV).

NEURON DENSITY

The density of neurons varies considerably between the different cortical areas of the cetacean brain, but also within individual layers, with the second layer accounting for about half of all the neurons in a given cortical block. Larger brains have more neurons in total than smaller brains, but they have a lower density of neurons. Neuron size does not directly relate to brain size, but there is a tendency in cetaceans and other mammals for neuron size to increase as brain size becomes larger.

An increase in brain size is also accompanied by an increase in the number of glial cells, which are non-neuronal cells that protect, nourish, and isolate neurons. Cetaceans have high numbers of glial cells but no more than would be expected for their brain size. Recent studies have shown that the organization of the cetacean cortex is comparable in complexity to that of land-based mammals and that the cetacean cortex has evolved an alternative way to generate complex behaviors.

The dolphin cortex

The cortex of the dolphin brain is extremely folded and composed of only five cortical layers. The neuronal organization is complex, comparable to that of other terrestrial mammals, and considered to be an evolutionary alternative to the generation of complex behaviors.

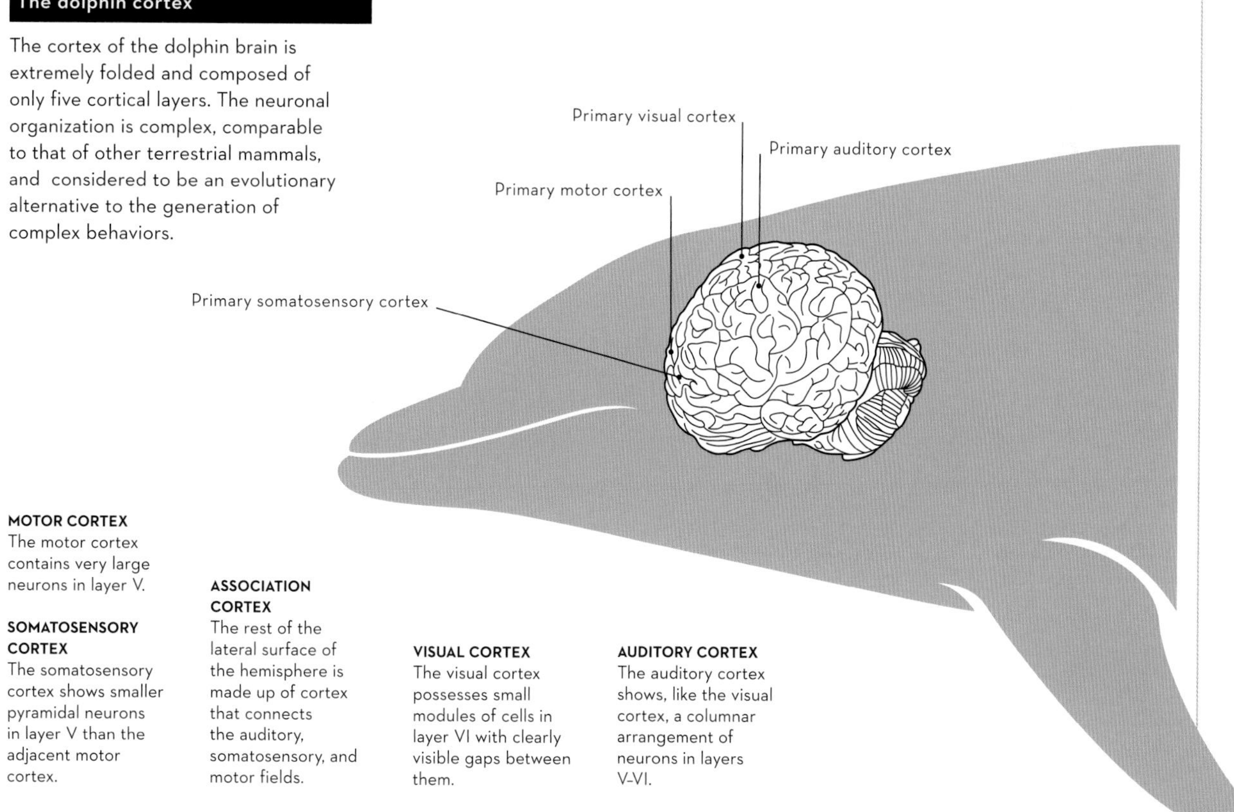

MOTOR CORTEX
The motor cortex contains very large neurons in layer V.

SOMATOSENSORY CORTEX
The somatosensory cortex shows smaller pyramidal neurons in layer V than the adjacent motor cortex.

ASSOCIATION CORTEX
The rest of the lateral surface of the hemisphere is made up of cortex that connects the auditory, somatosensory, and motor fields.

VISUAL CORTEX
The visual cortex possesses small modules of cells in layer VI with clearly visible gaps between them.

AUDITORY CORTEX
The auditory cortex shows, like the visual cortex, a columnar arrangement of neurons in layers V-VI.

COMPARING THE BRAINS OF ODONTOCETES & MYSTICETES

Although similar selective pressures shaped the brains of odontocetes and mysticetes during their evolution, several key differences exist between these two suborders in terms of external brain shape, size, and organization. First, there has been stronger selection on brain size in odontocetes than mysticetes. Second, the distinct socio-ecological strategies of odontocetes and mysticetes in relation to feeding, migration, and social structure and complexity have favored elaboration and regression of different brain regions. Finally, the variation in brain size and structure, even among closely related species, presents a considerable challenge for scientists who are interested in what social and ecological factors shape brain (and cognitive) evolution.

BRAIN STRUCTURE & CORTICAL ORGANIZATION

Although similar selective pressures shaped the brains of odontocetes and mysticetes during their evolution, several key differences exist between these two suborders in terms of external brain shape, size, and organization. First, there has been stronger selection on brain size in odontocetes than mysticetes. Second, the distinct socio-ecological strategies of odontocetes and mysticetes in relation to feeding, migration, and social structure and complexity have favored elaboration and regression of different brain regions. Finally, the variation in brain size and structure, even among closely related species, presents a considerable challenge for scientists who are interested in what social and ecological factors shape brain (and cognitive) evolution.

Large spindle neurons, known as VENs, are found in the fifth layer of the cortex in both odontocetes and mysticetes. These are similar in size, shape, and distribution to those present in the brains of humans and apes. VENs are also found in many other mammals including elephants and the closest living relative of cetaceans, the pygmy hippopotamus, but with very different patterns of distribution within the cortex. VENs have been suggested to play a role in autonomic regulation and in social, cognitive, and emotional networks in humans, apes, elephants, and cetaceans, but there is no direct evidence that they serve such functions in all species in which they occur. The different distribution of VENs in different species indicates, indeed, that VENs might be involved in other functions depending on their cortical distribution and that such neurons are probably involved in different networks in different species.

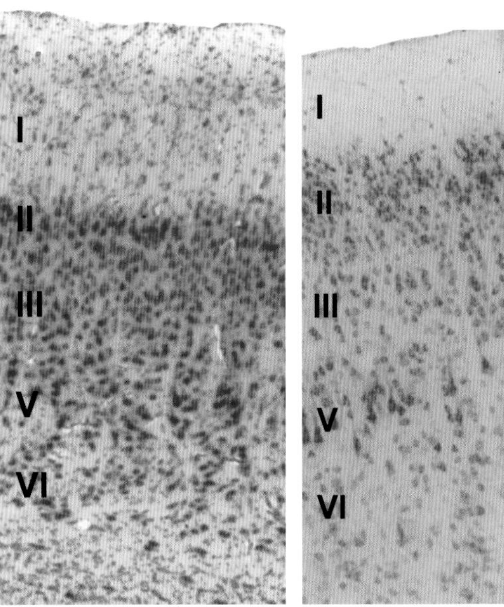

Above The auditory cortex of a bottlenose dolphin (left) and the motor cortex (right) of a beluga whale. The auditory cortex is the region of the cortex that processes auditory information and presents, microscopically, groups of neurons arranged in vertical columns through layers V and VI. The motor cortex is the region that controls the execution of movements. It is characterized microscopically by the presence of large neurons (pyramidal neurons) in layer V.

Comparing odontocete & mysticete brains

Major differences of brain organization between odontocetes and mysticetes are related to local neuronal densities across cortical areas and size of the neurons.

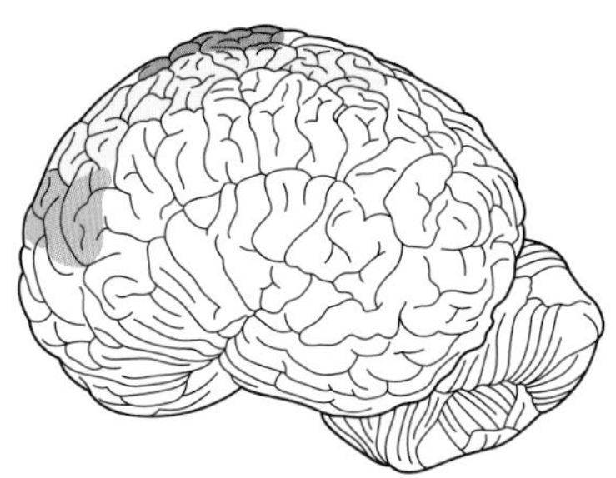

DOLPHIN (ODONTOCETE) LATERAL

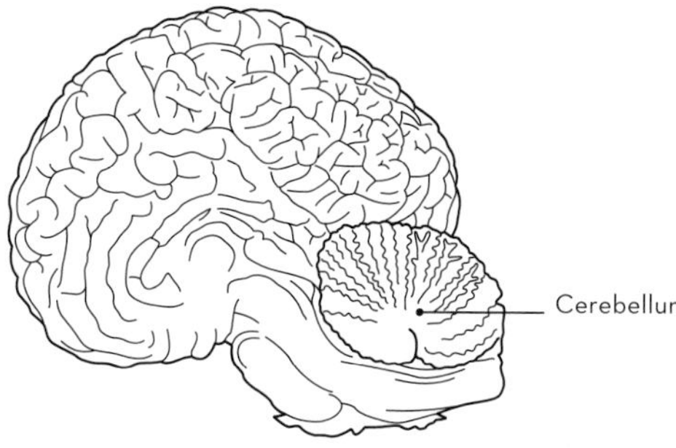

DOLPHIN MIDLINE

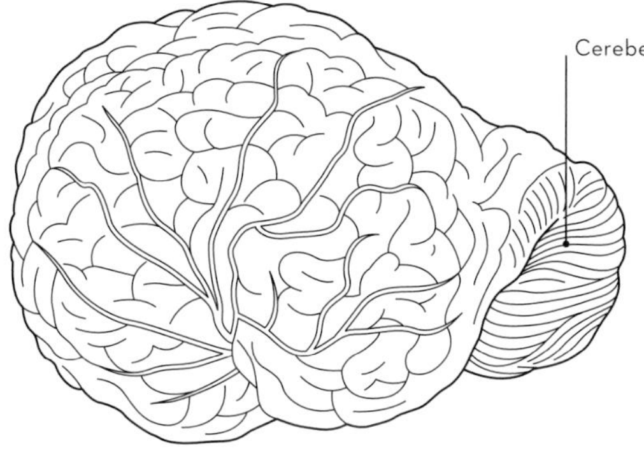

HUMPBACK WHALE (MYSTICETE) LATERAL

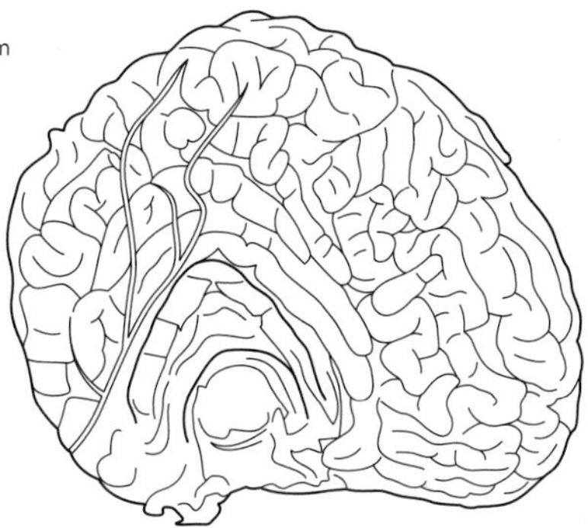

HUMPBACK WHALE MIDLINE

BRAIN SHAPE
The odontocete brain (top) is shorter and wider than the mysticete brain (above). This is caused by a more pronounced evolutionary process in odontocetes of brain shortening along the beak-fluke axis.

GYRAL PATTERN
Gyrification—the characteristic cortical folds—increases the area of the cortex and the space available for an increased number of neurons without necessarily increasing the size of the brain. This pattern of gyrification is less pronounced in mysticetes, possibly because their wider cortical layers create a thicker cortex with more space available for neurons.

CEREBELLUM
The mysticete cerebellum is rounded and smaller, relative to overall brain size, than the odontocete cerebellum, which appears flattened dorsoventrally as a consequence of the pronounced shortening along the beak-fluke axis.

HIPPOCAMPUS
The hippocampus is small across all cetacean species but even smaller in mysticetes. This might be because spatial cues are not particularly useful for migratory species traversing vast areas of ocean.

LAYERING
Layering patterns in the cortex are comparable in odontocetes and mysticetes. However, neuron size and density vary widely depending on brain mass. Even though mysticetes have larger brains in general than odontocetes, some species of odontocetes with exceptionally large brains, namely sperm whales and killer whales, also have brains with a lower density of larger neurons.

SPECIALIZATIONS
Odontocete brains include clusters of neurons in the second layer of the ventral portion of the insula (a region deep in the cerebral cortex), which is a cortical region involved in perception, emotion, and maintenance of internal environment. However, in the humpback whale and fin whale, both species of mysticetes, these clusters are extended farther to the occipital cortex and inferior temporal cortex. This peculiar distribution of clusters is restricted to mysticetes and may represent a cortical specialization. The functional significance of these clusters is unknown but they may be involved in optimizing connections between the hemispheres in large brains by reducing the energetic cost.

BRAIN MASS, BODY MASS & THE ENCEPHALIZATION QUOTIENT

Understanding how changes in brain size affect behavior and correlate with intelligence, if at all, is a difficult task. Mammalian adaptation to an aquatic environment is not unique to cetaceans, yet they include species with the largest brains of all mammals, both overall and relative to body size. Because neural tissue is expensive to support, it is thought that species with relatively large brains must be using them to support their survival and reproduction. While brains become larger with body size they do not scale together perfectly. Furthermore, many large animals manage just fine without particularly large brains (think of whale sharks or Cape buffalos). This raises the question of what exactly do cetaceans use their large brains for?

BRAIN MASS & BODY MASS

Odontocete brain masses vary from 7¾ oz (220 g) in the La Plata dolphin up to 17 lb 9 oz (8 kg) in the sperm whale, and mysticete brain masses range from 7 lb 9 oz (about 3.5 kg) in the blue whale up to 14 lb 12 oz (6.7 kg) in the fin whale. It's clear just by looking at these figures that a larger body does not necessarily correspond to a larger brain. The blue whale is the largest and heaviest animal on the planet, so we might expect its brain to be the largest in absolute size compared to other mammals.

To compare properly brain mass among different species we must take into account brain mass in relation to body mass. However, this isn't straightforward because brain mass does not increase linearly with body mass. If this were to be our measure then the largest mammalian brain would in fact belong to the small tree shrew, whose brain accounts for 10 percent of its body mass. The human brain accounts for only 2 percent of body mass, so clearly the link between brain mass and intelligence is at best inconclusive.

THE ENCEPHALIZATION QUOTIENT

How, then, can scientists measure the intelligence of a species while taking all this into account? The main problem is the impossibility of producing a universal definition of "intelligence" and, as a consequence, a universal way of measuring it. Intelligence, when defined as mental and behavioral flexibility, has evolved in many groups of mammals, yet humans are considered to have surpassed all other species in this regard. Most definitions begin with the assumption that humans are at the pinnacle.

A simpler task than defining and measuring animal intelligence is to measure brain size. One method scientists use is to compare a species' brain mass with the brain mass expected for a given body mass. This value is known as the encephalization quotient (EQ). An EQ equal to 1 means that the brain has the expected size; an EQ larger than 1 means that the brain is larger than expected; and an EQ smaller than 1 means that the brain is smaller than expected. In theory the greater the EQ value, the greater the intelligence of that species. There is a large variation in EQ values among cetaceans. During cetacean evolution, there was a substantial increase in odontocete encephalization that resulted in some species of porpoises and dolphins having EQ values second only to humans. On the other hand, mysticete encephalization decreased due to a disproportionate increase in body size that did not require a matching increase in brain size. As a result, values of EQ within mysticetes are generally below 1 and do not realistically reflect the evolutionary process of brain enlargement in cetaceans as a whole. This also holds true for some odontocete species with particularly large bodies, such as the sperm whale. So while EQ can predict encephalization in several groups, it is unclear whether EQ can be used as a proxy for intelligence.

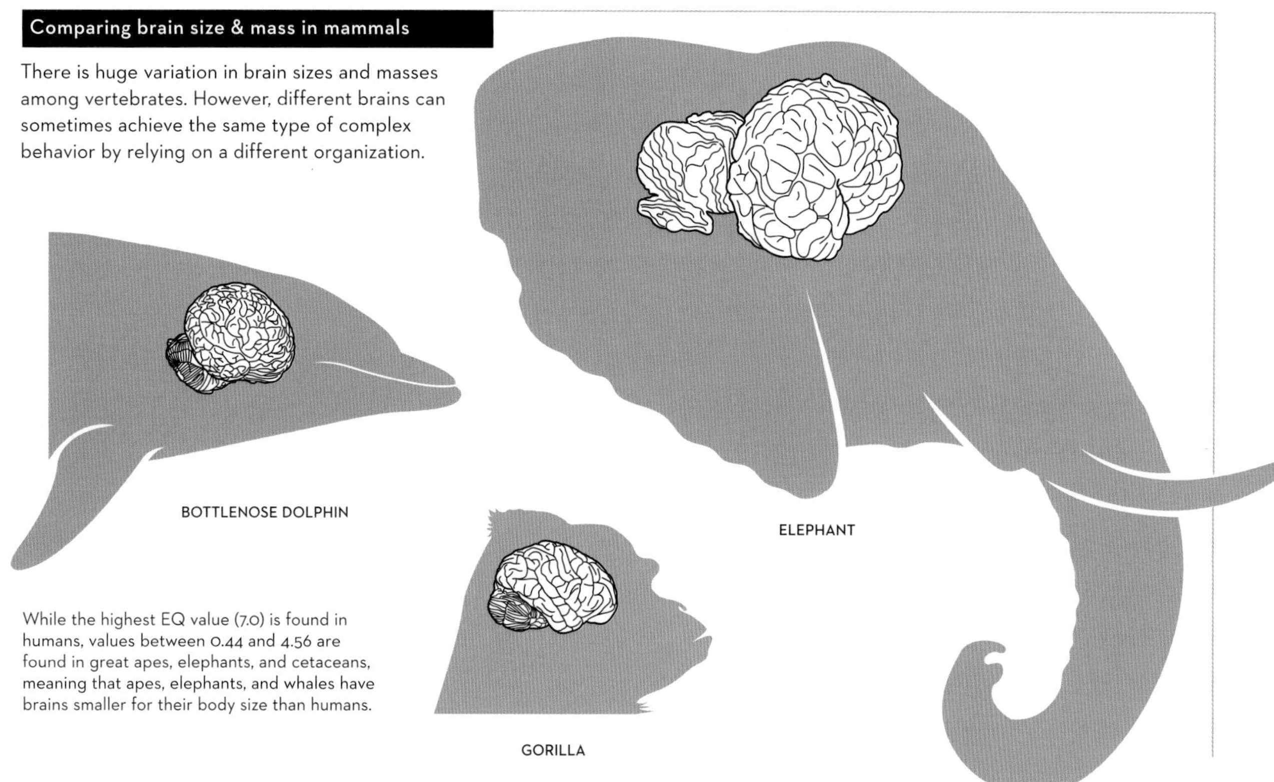

Comparing brain size & mass in mammals
There is huge variation in brain sizes and masses among vertebrates. However, different brains can sometimes achieve the same type of complex behavior by relying on a different organization.
BOTTLENOSE DOLPHIN
ELEPHANT
GORILLA

While the highest EQ value (7.0) is found in humans, values between 0.44 and 4.56 are found in great apes, elephants, and cetaceans, meaning that apes, elephants, and whales have brains smaller for their body size than humans.

WHERE DO WE GO FROM HERE?

Current knowledge of brain anatomy and function in cetaceans is based on a handful of species from which brain specimens have been relatively easy to obtain and access. In view of the remarkable diversity of both suborders (odontocetes and mysticetes) these observations, while important as a baseline, cannot be considered definitive and, most certainly, do not reflect orders and species differences adequately. In fact most information about the cetacean brain anatomy is derived from opportunistic observation of isolated specimens while a more substantial amount of data is gathered from a few delphinids, particularly the bottlenose dolphin.

Future research must endeavor to document the fine structure of the brain of species that are less common or rare. Studies on a large comparative scale, within cetaceans, are essential to reveal differences in the organization of cortical regions that might be evolved in response to specific adaptive processes in this group of mammals fully adapted to life in water. To understand why some cetaceans evolved bigger brains or ones with specific characteristics, we need more comprehensive data from all odontocete and mysticete species. Similarly, data on social behavior, social structure, and diet can help us understand broad evolutionary patterns, just as primate studies found simpler social organization and feeding specializations were associated with smaller brains than more complex. In this context, it would be important to compare brain structure in cetacean species known to be more social (bottlenose and spinner dolphins, pilot whales) to more solitary species—Commerson's dolphin for example. Comparisons of brain specialization in estuarine and river dolphins with their saltwater relatives would be invaluable, especially given the variation in brain size among cetaceans and evidence for positive selection on neuronal traits in bottlenose dolphins.

Understanding the organization of such a complex and large brain as that of cetaceans is fundamental to elucidating the more general rules that drove the evolution of the mammalian brain. More comprehensive neuroanatomical and behavioral data on a greater number of species are needed to understand how selective pressures acted during evolution to result in the variety in brain size and organization observed among cetaceans. It will ultimately contribute to a better understanding of these species' cognitive and behavioral adaptations to their natural habitat.

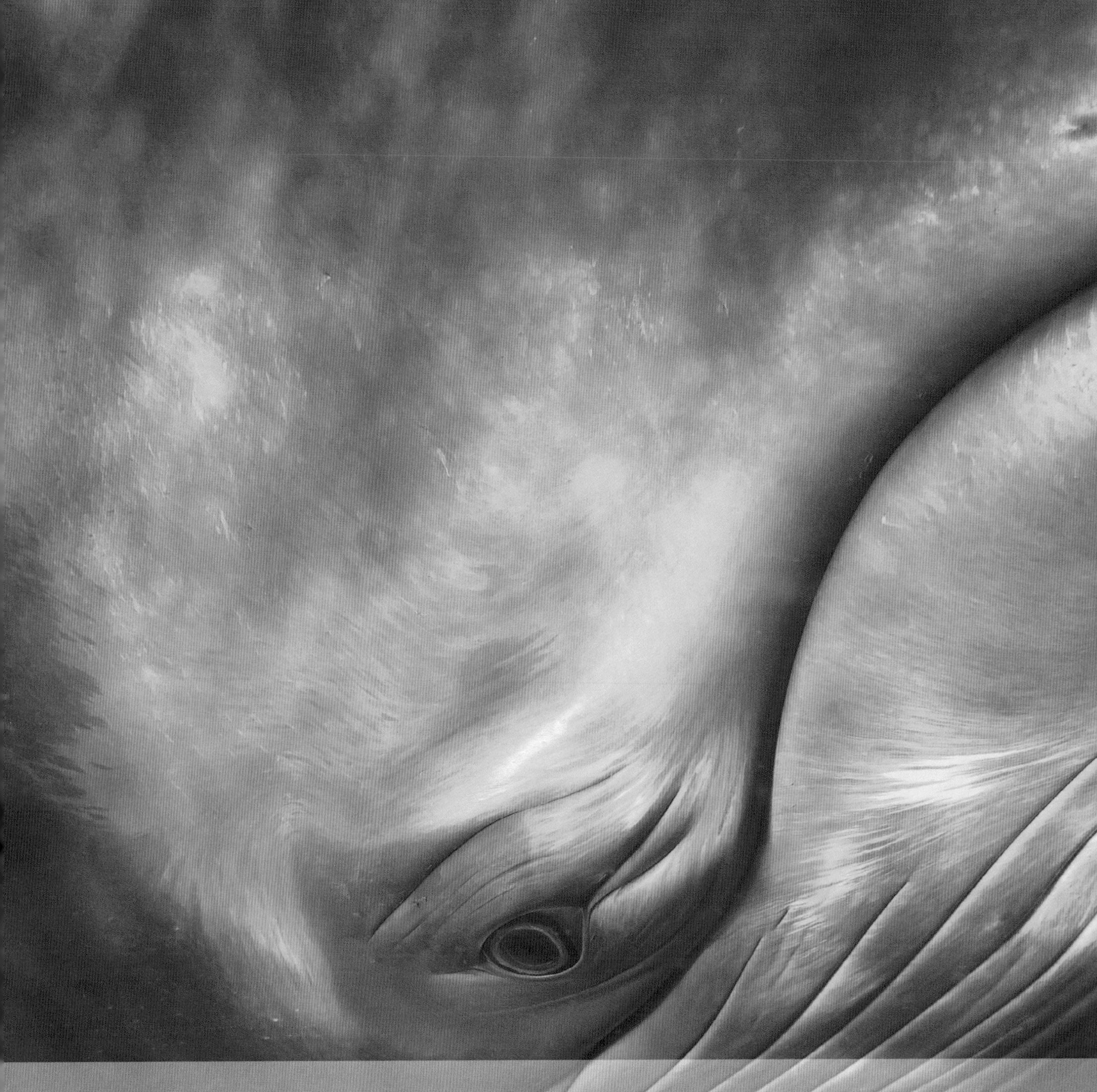

3 COGNITION

Heidi E. Harley

Untitled, Dwarf Minke Whale
Great Barrier Reef, Australia, 2009

"This is another photograph of the whale we named Ella as I waited for her to approach closer for the portrait session to begin. Ella's curiosity ultimately allowed me to create my largest, life-size, full body composite photograph of a whale measuring two by nine meters."
—Bryant Austin

KNOWING WHAT DOLPHINS KNOW

The sun rises on the Common Era as a young Roman boy walks on a narrow isthmus between the Gulf of Naples and Lucrine Lake. He calls "Simo! Simo!" and wades in as the familiar sleek dorsal fin breaks the surface near the shore. The saltwater lifts the boy up as the two mammals greet each other and the boy climbs onto the dolphin's back for this daily ride across the inlet to the Puteoli school.

Do you doubt the veracity of this story? Readers of the time likely did not doubt when Pliny the Elder published this account before his death in 79 CE, not only because it was well known in the Baiae region but also because it followed a long line of similar stories—from Odysseus's son being saved by a dolphin to Greek boys and their dolphin companions to Aristotle's more scientific observations of dolphins' mammalian attributes and tendency to support flagging fellow dolphins.

For many this story likely seems fanciful, but even as you read this sentence there may very well be a person in Ireland, the United States, Brazil, or Australia interacting with a wild dolphin that solicited the encounter. Why? We do not know. In some cases the dolphins are clearly seeking—and getting—food from the humans, but this does not occur in all cases. What we do know is that this intertwined history between our species has inspired Greek coins and art, with boys on dolphins, stories of gods and goddesses who revere dolphins or even are dolphins (for example, the perhaps surprising identification of a desert-dwelling people, the Nabataeans, with a dolphin goddess), and our continued interest in dolphins and the mystery of their minds. Their intelligence and skill have also inspired wild speculations and assertions about their abilities and origins (did they appear on Earth directly from outer space?) as well as scientific study to ascertain their earthbound strengths and capacities. Here we explore some of the cognitive powers of bottlenose dolphins (the most commonly studied dolphin species—like TV/movie star Flipper) and the scientific methods we use to uncover those powers.

STUDYING COGNITION

How does one explore the mind of a dolphin? By asking and answering questions with the tools one can muster. One such tool is a matching task. It works as its name implies: we present a sample stimulus to the dolphin and the dolphin must choose its match from among a set of alternatives in order to receive some desired outcome. For example, let's say we want to find out whether dolphins can tell the difference between a scallop shell and a mussel shell. To ask this question, we could present the dolphin with a scallop shell, take it away, and then present the dolphin with a similar scallop shell and a mussel shell. If the dolphin chose the scallop shell, we could indicate to him that he'd chosen correctly by giving him a fish. If he chose the mussel shell, we could try the trial over again until he got it right. Ultimately, we would want to test him in a situation in which there were many controls so that we could have faith in our interpretation of his answer. For example, we would want to have an equal number of trials with a scallop as the sample and a mussel as the sample so that he didn't get more fish for one shell than another, present the choice shells such that the right answer showed up an equal number of times on the right and on the left in the alternative array so that placement didn't provide the answer, and make sure that the person or device presenting the alternatives did not give anything away about the correct choice.

Above A replica of an ancient Greek coin dating from 380–345 BCE depicting a dolphin carrying a boy on its back.

Above Detail from the mosaic floor, dating from the fourth century CE, of the basilica in Aquileia in north east Italy. Aquileia, once one of the largest and richest Mediterranean cities within the Roman Empire, is now a UNESCO World Heritage Site. The theme of the mosaic is an example of a "Bible of the poor" (*Biblia pauperum*), intended for those who could neither read nor write as a means of understanding the sacred story through images. The magnificent mosaic is richly depicted with saltwater animals including what appears to be a dolphin in this detail.

It turns out that when people are doing the presenting, the best way to make sure they don't give cues about the correct choice is to ensure that they don't know which choice is right. Animals can be quite good at picking up on knowledge that another animal (including a human animal) is not intentionally trying to convey but conveys nonetheless—like leaning slightly toward the mussel shell when it's the match.

READING CUES

Perhaps from the glimpse into the controls listed above, you have already gleaned that the matching task has been well developed in laboratory research with many species including birds and human/nonhuman primates. One famous example related to the controls necessary for studying cognition in animals centered on an early twentieth-century horse, Clever Hans, who was purported to be able to answer complex math problems by stamping his hoof. Careful experimentation revealed that he was actually reading very subtle—and unintentional—cues produced by his trainer who himself knew the answers to the problems. So, although Hans couldn't do math, he was actually an expert at reading humans. Because appropriate controls rule out alternative explanations for an animal's behavior and almost always require many trials, cognitive studies are conducted with animals who are consistently accessible to humans and thus live in human care. Doing cognitive studies with wild dolphins is rare because the origins of a behavior are typically central to interpreting it through a cognitive lens and because one has to interact with the same animals very often over long periods, a situation that is typically not in the best interests of wild animals even when it is possible.

Although there are many ways to address cognitive questions, the matching task is often enlisted due to its versatility in addressing all kinds of different questions. In the scallop/mussel shell example, there was a small delay between the presentation of the sample and the presentation of the choices. If we wanted to study memory, we could increase that delay to determine how long the dolphin could remember the answer. If we wanted to know if he could discriminate more shells, we could increase the size of the choice array from two kinds of shells to four or six. If we wanted to know if he could use echolocation/biosonar to recognize the objects, we could ask the dolphin to wear a blindfold and to do the task with echolocation alone.

LEARNING CONCEPTS

The matching task also offers another opportunity: it can be used to ask questions about concepts. One such concept is the matching concept itself. Can a dolphin learn the matching concept? Here you might want to suggest that by doing the scallop/mussel discrimination, this dolphin has already shown he can—but at this point we know that he can do the task with just these two shells. To find out if he's using a general rule, we'd want to change the stimuli (that is, the objects we present) to see if he'll continue to match with new stimuli.

Animals—including humans—often learn a simpler rule when there are fewer stimuli, but they are more likely to learn a general rule after more exposure to many stimuli. For example, when toddlers are learning language they often learn the irregular past tenses of the verbs they hear the most often early on, such as "Kitty ate da fish." As they become more expert with language, the sentence changes in an unexpected way, "Kitty eated da fish." Can we call the progression from a well-constructed sentence to an ungrammatical one progress? You bet. The second sentence includes a sophisticated error. In the first sentence, the child is imitating an adult; she has memorized the verb "ate." In the second sentence, the child is almost certainly not imitating an adult because it is unlikely that an adult would have produced the word "eated." Rather, the error shows us that the child has learned an important rule, adding "-ed" to verbs to create the past tense, and then has "transferred" that knowledge to a verb in a form she's never heard uttered. In a similar vein, if we want to verify that a dolphin knows the matching concept, we test the dolphin in transfer tests with new stimuli to see if he can immediately match new objects rather than needing the cues that familiar stimuli might provide.

This idea of learning about generalized concepts (choose the matching object) versus learning only about specific stimulus-response pairings (mussel shell choose mussel shell) will reappear throughout this chapter as we look deeper into the dolphin mind and how we study it.

Opposite A young female bottlenose dolphin photographed off the Caribbean island of Curaçao, displaying what appears to be playful behavior, balancing a yellow clam shell on her rostrum.

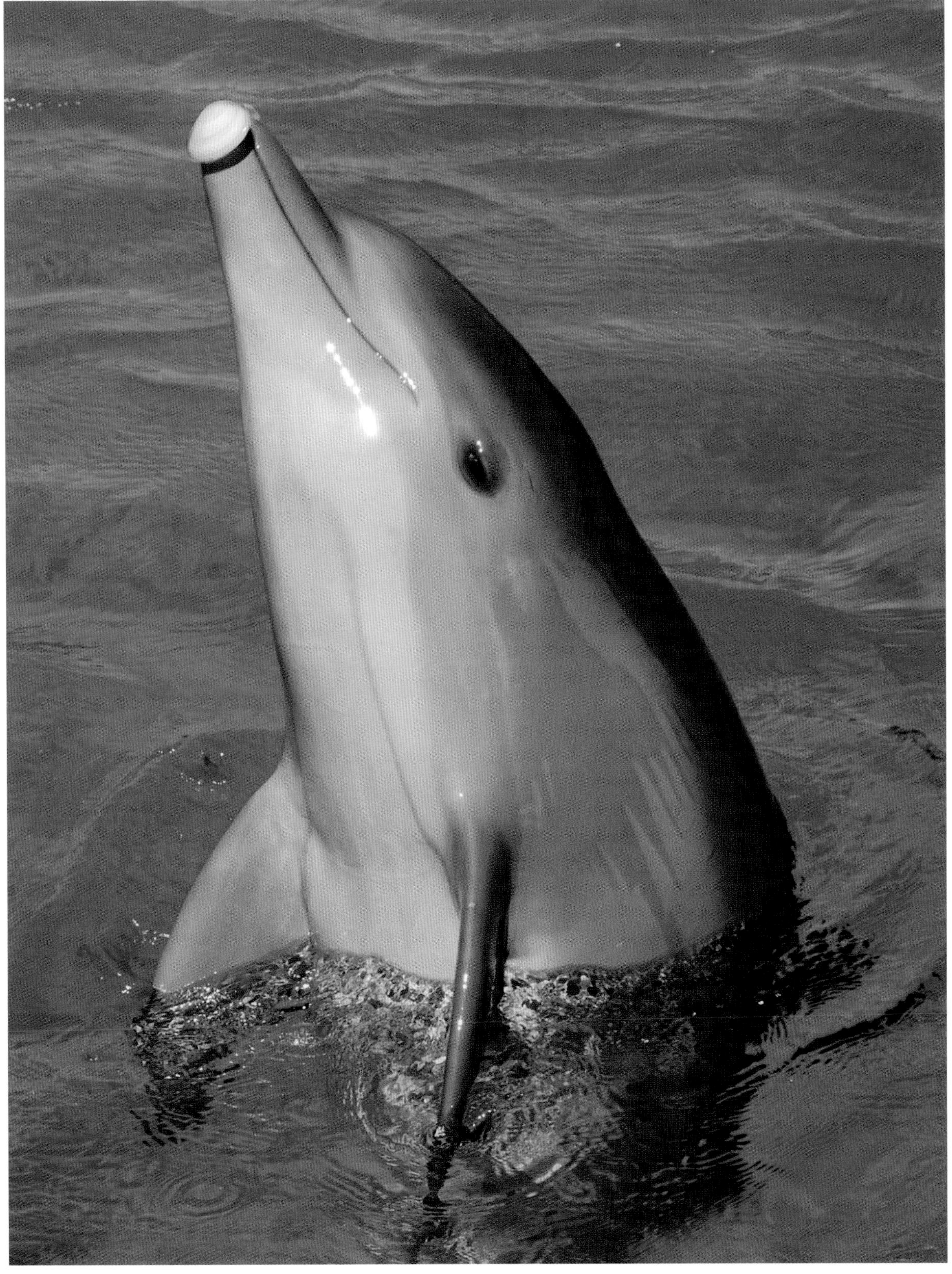

DOLPHINS KNOWING HUMANS

Just as the mystery of the dolphin has inspired stories of interspecies friendships through millennia, so has it also inspired fiction writers and researchers to take a King Solomon's Ring approach—they have imagined or created systems that might enable interspecies communication, providing real windows into another species' mind. Have they succeeded with the dolphin? After reading this section, you can decide for yourself.

One of the earliest human-dolphin communication projects occurred at a houseboat and lagoon situated within an island featured in the opening credits of the 1960s sitcom *Gilligan's Island*. At this Coconut Island lab located off Hawaii's Oahu, Wayne Batteau and Peter Markey worked with two dolphins, male Maui and female Puka. The US Navy supported the two researchers to create devices that translated human vowel sounds into whistles and vice versa. Maui began the project alone at the Point Mugu Cetacean Facility in California where he purportedly learned to imitate five sounds and to perform five different behaviors to those sounds: BIP hit a ball with the pectoral fin, BAIEP go through a hoop, BEIAP raise the flukes, BAEP click, UWEIAP roll over. After Maui moved to Hawaii and Puka joined him, Maui's repertoire expanded a little and Puka was exposed to the same training. The study lasted for three years, and its outcome was difficult to interpret. There was indeed a device that turned vowels into whistles, and there was a way for people to see the dolphins' whistles visually (rather than having them translated into human words). However, the stability of the dolphins' behavioral responses, vocal imitations, and ability to learn from each other was not clearly enough established to inspire the confidence of either Batteau and Markey or an outside reader of the resulting technical report. On the other hand, the old black and white photos of humans and dolphins working together on this idyllic Hawaiian island did evoke some sense of the wave of optimism on which this venture was riding.

A good deal of the energy for human-dolphin communication during this era came from the scientific and popular writings of John Lilly, an MD neuroscientist who performed invasive brain surgeries on dolphins, carefully recorded their vocalizations, and even set up a house in the US Virgin Islands in which a woman, Margaret Howe, and a dolphin named Peter lived together for a time (see pages 44–5). At first, Dr. Lilly published scientific journal articles in which he reported interesting findings. For example, he and Alice Miller reported some of the earliest information about dolphin vocalizations in the journal *Science*, that is, that dolphins can produce clicks and whistles both separately and at the same time, and that their vocalizations to each other are often antiphonal, meaning that they take turns vocalizing like a conversation rather than overlaying their sounds on top of each other like the chorusing of frogs or crickets. Lilly's dolphins also listened to a human model produce a series (1 to 10) of syllables from a group of 198 syllables, and Lilly recorded the dolphins' "imitations" of the syllables; he discovered that the dolphins usually produced the same number of syllables as the human models. Ultimately, Lilly's claims about dolphin communication, language, and intelligence far exceeded the available data. Lilly's assertions about their linguistic prowess were premature then, and still are today (see Chapter 4).

Lilly's claims and Batteau and Markey's work did bear scientific fruit in Lou Herman's lab at the University of Hawaii. Herman's dolphin work began when he discovered he was allergic to the rats in his current lab and one of his students, Frank Beach III, suggested that they do a class project with dolphins at Sea Life Park, an oceanarium started by entrepreneur Tap Pryor that also included a research branch, the Oceanic Institute. The project was so successful that it led to a publication about dolphin learning in a scientific journal and, ultimately, to the formation of a dolphin lab, the Kewalo Basin Marine Mammal Laboratory (KBMML—typically referred to as Kih-bih-mul by its community). This facility became a mecca for the study of dolphin learning and cognition for several decades. The lab was in part a direct legacy of Wayne Batteau's work in that

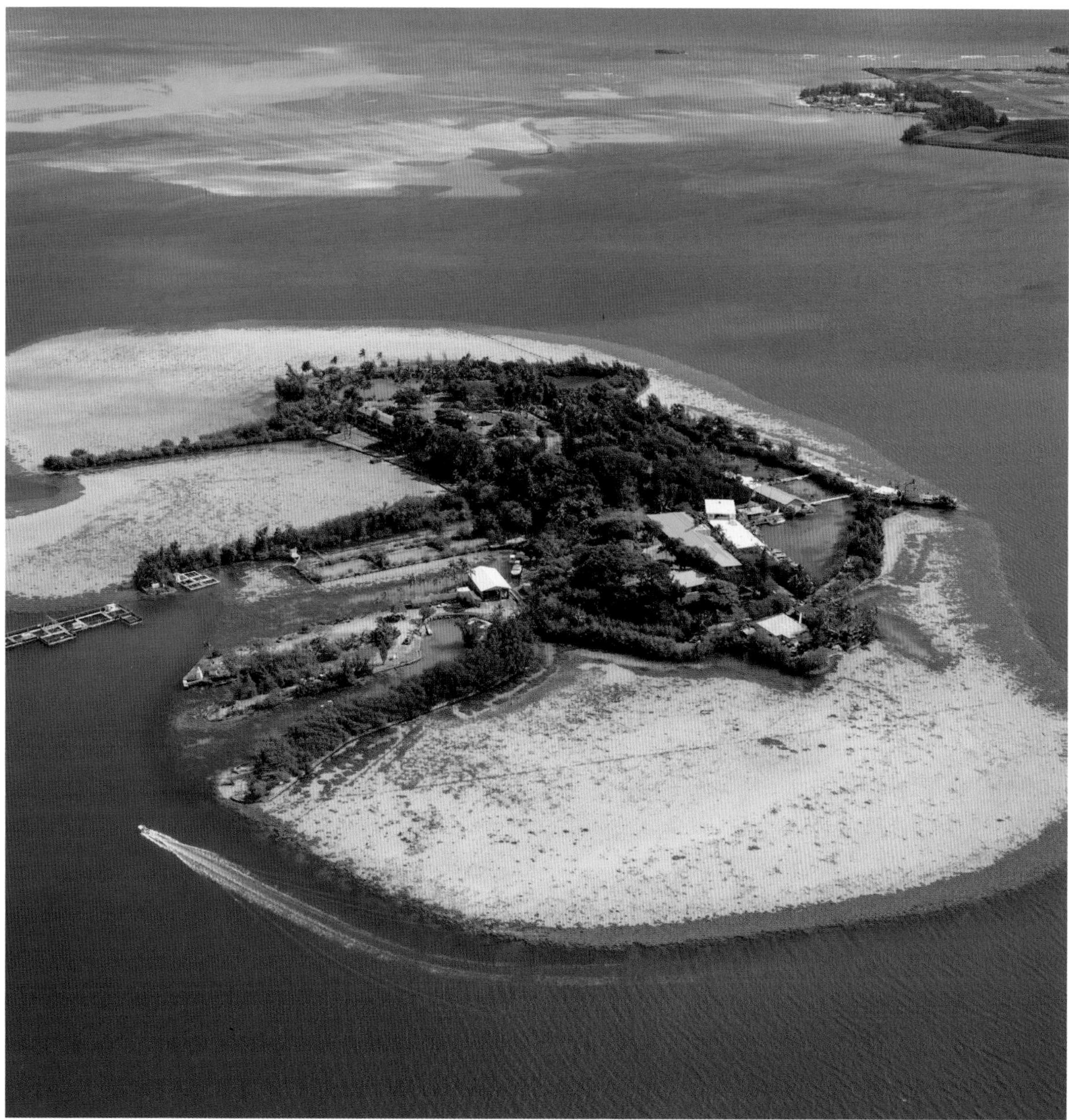

Above Coconut Island, Kaneohe Bay, Hawaii. The island was the idyllic location for a Washington-funded research laboratory, run by Wayne Batteau and Peter Markey, in their pursuit of a two-way communication system between humans and dolphins.

one of the dolphins engaged in the early work was Puka of Coconut Island fame. The other long-term resident female was Kea. Although Puka and Kea taught us a great deal about dolphins' abilities to process visual and acoustic information, their participation in human-dolphin communication ended early when they were dolphin-napped by a couple of workers at Herman's lab and dumped from a pick-up truck into the Pacific—never to be reclaimed. (Captive-raised dolphins have not survived releases into the wild.) However, Herman continued his communication project with two more females, Phoenix (who helped the lab rise from the ashes of Puka and Kea's loss) and Akeakamai (Hawaiian for "lover of wisdom"), or Ake for short, who arrived at the lab in fall 1978.

AN ADVENTUROUS MIND:
JOHN CUNNINGHAM LILLY

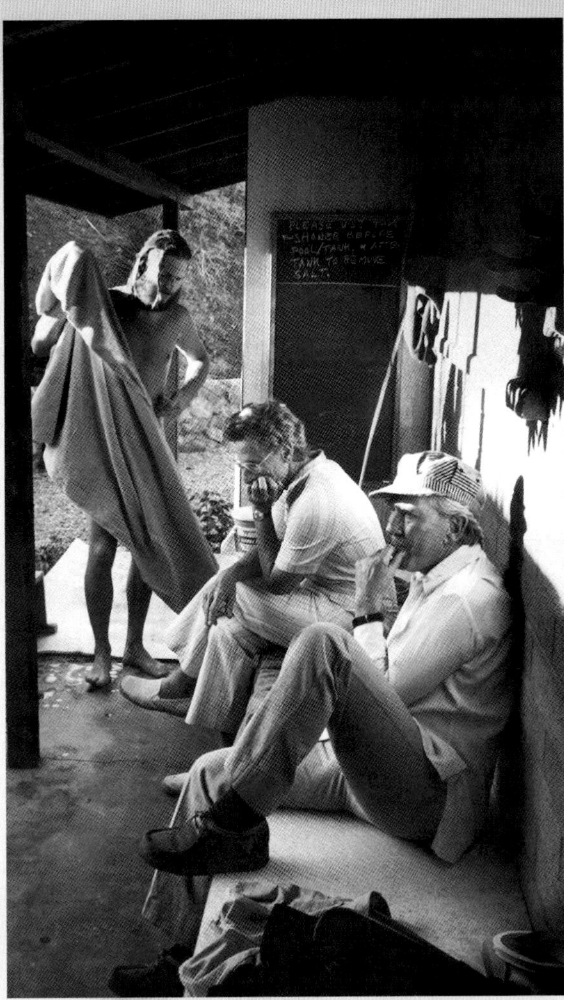

Above Dr. John Cunningham Lilly (center) sitting with the actor Burgess Meredith (right) outside the isolation tank as actor Jeff Bridges (left) towels off after his turn inside.

JCL RESUME I: *Floating in isolation tanks, tripping on LSD, promoting sexual encounters with dolphins, investigating himself to unlock the universe, communicating with aliens.*

JCL RESUME II: *Bachelor's degree from CalTech, MD from Penn, Principal Investigator with major science funding agencies (for example, National Institute of Mental Health, Office of Naval Research, National Science Foundation), neuroscience pioneer.*

As these dual résumés suggest, John Lilly cavorted through his time with a fine mind, good training, and a relish for discovery and adventure that pulled him from one unknown to another. And what a time it was. From 1915 Minnesota to 2001 Los Angeles, Lilly not only witnessed all the changes of his generation—from the devastation of the World War I flu epidemic to the sophistication of the twenty-first-century Center for Disease Control, from horse and carriage to space travel, from the telegraph to the internet, from the discovery of penicillin to the worry of antibiotic resistant bacteria, from no US women voters to the 67 US federal Congress women of 2001—but he also played a public role that pushed every boundary he encountered. He tuned in, but he certainly didn't drop out. Rather, he led a host of followers into his dreams—literally. If you want to go there in fiction form, read Chayefsky's novel *Altered States* or watch the movie written by the same author, both based on Lilly's self-investigations into taking psychotropic drugs and floating in isolation tanks. Lilly identifies his interest in using floating to investigate the mind as the underpinning of his initial focus on dolphins, potential kindred saltwater floaters.

As a neuroscientist, Lilly studied dolphins through the dissection of dead brains and the electrical stimulation of live ones, and he decided that the large size of the dolphin brain coupled with this species' complex vocalizations did indeed make the bottlenose dolphin a kindred spirit. What do you do when you've found a kindred spirit? Spend time together, talk, hang out. And that's what he did—or at least what he arranged. He worked with a practical and passionate woman, Margaret Howe, to design and test human-dolphin cohabitation in a house in St. Thomas in the US Virgin Islands. There, at the Dolphin Point Laboratory, Howe tested the waters by living with dolphin Pam for a week, doing some evaluating and remodeling, and then living with dolphin Peter for two and a half months in a house with a couple of flooded rooms. Howe spent almost all of her time with Peter. She slept on a large, foam mattress surrounded by shower curtains, and Peter usually slept beside the bed. She cooked, wrote, talked on the phone, brushed her teeth—did everything one does to live—with Peter, and they spent the days interacting with each other including participating in formal lessons. The original goals of the work were to investigate the dolphin's learning ability, to teach the dolphin to speak and understand English, and to find out how to improve the human-dolphin cohabitation environment. Howe kept a journal in which she included not only daily activities and pragmatic discoveries but also her own feelings of occasional loneliness as well as her evaluation of Peter's cognitive and emotional states. She also wrote about the mutual trust she felt that she and Peter had established through intimate contact: over the ten weeks, Peter taught Howe to trust him when he gently ran his teeth up and down her legs in what she perceived to be a form of courting, and he learned to trust her to use her hands or feet to bring him to orgasm.

Lilly and Howe's analysis of this unique venture used both scientific and intuitive methods, a match that can be very fruitful scientifically if one follows up the intuitive hypotheses with experiments but that was a path that Lilly chose not to take. Instead, he followed the pathways of his mind and took the public with him. He wrote/co-wrote more than a dozen books about mind—his own and his extrapolations to others' minds—and literally covered life, the universe, and everything including his representation of God, a hierarchically organized "guiding intelligence in the universe" through a Cosmic Coincidence Control Center of which Earth's part is ECCO, the Earth Coincidence Control Office. His writings were a powerful combination of confident assertions backed up by some scientifically sound reasoning and some remarkably exaggerated interpretations (from a scientific perspective) of what he had experienced.

Perhaps at this point you are wondering how to make sense of Lilly and his findings? Join the club. On the one hand, Lilly's work helped inspire generations of scientists to study dolphins—many of the cetacean scientists I know were drawn into the field by John Lilly. Hallmarks of the scientific process are discovery and playing with ideas, and Lilly had the courage, creativity, and intelligence to engage in these activities in spades. On the other hand, his charismatic persona drew a portrait of dolphins beyond anything we've managed to unearth to date, and most of us in this field find ourselves having to fight through this mystique to show people some of the amazing things that we actually do know real live dolphins do. Do dolphins have a sixth sense? Yes—echolocation! Can dolphins read minds? Yes—in the way that the famous horse Clever Hans (see page 40) could read small human movements, dolphins can pick up on human gestures and points, orientation and attention. Can dolphins communicate with each other? You bet—through body postures and whistles and bursts of clicks. They are fascinating creatures with remarkable acoustic abilities that developed in a marine environment quite different from our own. In that sense they are even aliens. When Lilly turned to dolphins, "for their minds in the water," it seems he jumped into the deep end—and the dolphins took him farther offshore rather than bringing him back to solid ground. The charismatic leader read his own charisma in their physiologically determined "smiles."

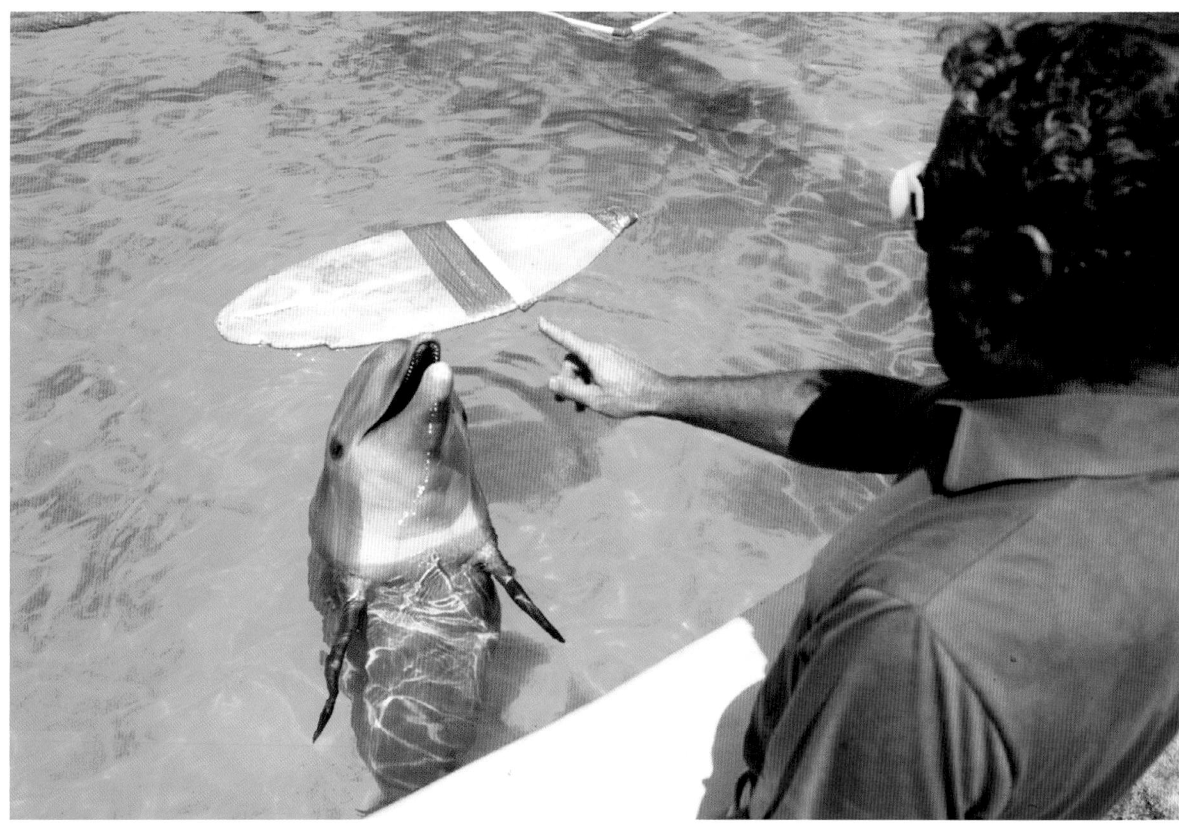

Above Lou Herman, who spent most of his professional life exploring the ways in which dolphins communicate. He tested dolphins' ability to respond to physical cues, such as hand gestures, described as a "dolphinized" version of American Sign Language. His work was funded for several years by the US Navy in hopes that inter-species communication might lead to dolphins assisting humans in specialized operations.

PHOENIX & AKEAKAMAI

Around the time of Phoenix and Ake's (pronounced Uh-K) arrival at KBMML, a scientific upheaval was occurring in animal language research, a field in which animals are exposed to two-way human-animal communication systems. For example, David Premack (University of Pennsylvania) had a talented chimpanzee named Sarah who learned to communicate using plastic icons, while Penny Patterson's Koko (a lowland gorilla) produced American Sign Language words at the San Francisco Zoo near Patterson's institution, Stanford University. Although the world was wowed by these studies, in 1979 Herb Terrace, a professor at Columbia University in New York, published an article in *Science* in which he argued that based on his findings with his own hand-signing chimpanzee, Nim Chimpsky, these animals were mostly only imitating and producing trained hand-signs in order to get food rather than using "language" in any meaningful way. Terrace's strong stance made skeptics of funders and the field slowed down. Terrace's work with Nim had its own set of problems, but the scientists doing similar work responded in the standard ways that science progresses: with critical analyzes comparing the methodological differences of Terrace's work with others, with analyses of the data to show other valid interpretations of the findings, and with changes to their methods to strengthen their work and to answer new questions.

Within this milieu, Herman chose to focus on a more one-sided human-dolphin communication system based on our knowledge of early language learning in humans. With human infants, comprehension of speech occurs before they begin talking, and it can be easier to measure what they understand compared to what they can produce because one can often measure their understanding by watching their behavior in response to commands. For example, children likely understand the word for shoes if you ask them to find their shoes and they bring them to you. Based on this logic, Herman decided to study the dolphin's comprehension of a human-created communication system rather than providing a system that they could both comprehend and produce. He wanted to find out if dolphins could learn some of the categories humans have in their languages (nouns, verbs, adjectives) and if the roles of these items could change when the sequence in which they were presented changed as it does in human languages: Tarzan and Jane climb vines versus Vines climb Tarzan and Jane.

Herman and his colleagues' work with Phoenix and Ake is the most intricate and rigorously analyzed human-dolphin communication project conducted to date. The two dolphins engaged in different systems. Phoenix learned an acoustic system based on computer-generated whistles, and Ake responded to human arm and hand gestures. Each system included items that stood for objects (for example, "Frisbee"), actions (for example, "fetch"), or modifiers (for example, "left" and "right"), and had its own order to test the flexibility of the dolphin's ability to process sequence in relation to meaning. Phoenix's system used a straight left-to-right sequence so that she could process items in the order they appeared: "Frisbee Fetch Ball" led her to get the Frisbee and take it to the ball. In contrast, Ake's system was inverted: "Frisbee Ball Fetch" led her to take the ball to the Frisbee. All utterances were commands from human to dolphin and were generally (except for some special tests) two to five items long. The controls were careful: trainers wore opaque eye coverings over their eyes so that they could not cue the dolphins about the locations of objects, and observers who did not know the commands that were given interpreted the dolphins' responses by generating the command that the dolphins had likely received. For instance, if Phoenix placed a sunken (as opposed to a floating) hoop into a basket the observer would say "Bottom Hoop in Basket" and Phoenix would only receive a fish if this were indeed the command given.

KNOWLEDGE TRANSFERENCE

One of the greatest strengths of Phoenix and Ake's systems was the opportunity to determine what the dolphins knew about each aspect of these imperative utterances both because the systems made it easy to generate specific commands with new combinations with which the dolphins had had no previous experience (i.e., one could look for transfer to new stimuli in order to determine whether or not they had learned a general concept about sequence and meaning) and because their responses were public and interpretable within the system's rules. For example, although Ake may have responded to "Frisbee Ball Fetch" many times during training and thus had been reinforced for this specific action (taking the Frisbee to the ball after seeing these same three gestures in this order over and over again), she could be tested with a novel command such as "Person Ball Fetch" (take the ball to the person in the water) that she had not experienced earlier and so had never been trained specifically to do. If she responded correctly to enough new commands, one could infer that she implicitly understood that "Person" could be an indirect object even though it may never have occurred as one before in her experience.

People are good at this skill; for example, you can likely interpret the sentence "Put the green beejbub in the soup." Although you do not know what a "beejbub" is, you do know that it's a direct object that goes in soup (unappetizing as it may sound). Similarly, Phoenix and Ake were good at this kind of transfer of knowledge; through their answers they showed that they could treat objects that had never appeared in a specific slot in a command as they should be treated (for example, as indirect objects or direct objects). They were also able to understand new varieties of sequences. For example, although Ake had only experienced a modifier in a one-object command before (for example, "Right Surfboard Over," jump over the surfboard on the right rather than the one on the left), she could interpret it correctly in a two-object command ("Right Water Hoop Fetch," take the hoop to the stream of water on the right versus the left) without extra training thus indicating that she understood the overall application of modifiers and object placement as dictated by the sequence of the items in the utterance. She also applied the object gestures to new similar objects such as Frisbees, balls, and surfboards of different sizes, shades, and overall configurations. Herman and colleagues continued to use Ake's system to ask novel questions for many years beyond its initial instantiation (Phoenix's system eventually became defunct due to technical difficulties generating the acoustic commands and so Phoenix began to participate in other kinds of research projects). For example, they installed Yes and No paddles in the tank and questioned Ake about the presence (Yes) and absence (No) of objects, and she responded appropriately.

As you likely know, dolphins are very good at reading people and their gestures. Proof of this capacity occurs in every oceanarium in the world. However, Phoenix and Ake's actions indicated something new. These two dolphins showed that their species could interpret sequence-based communication systems in which acoustic and gestural items could be differentially interpreted based on their positions within regulated strings. The dolphins were not trained to perform a specific action for each single gesture they fielded as is common in oceanaria; rather, Phoenix and Ake learned a flexible system that allowed them to immediately interpret novel commands. This flexibility is a form of concept rule-based learning.

TWO-WAY COMMUNICATION

Interesting as Herman's work was, researchers—and the world—still wanted more. They wanted a two-way communication system in which dolphins had more freedom to communicate back. Although Phoenix and Ake did communicate their understanding through their actions, they lacked the power to direct humans in the way that humans directed them. To fill that gap, several researchers tried providing the dolphins with keyboards.

In the 1980s at an oceanarium in Vallejo, California (then known as Marine World/Africa USA), Diana Reiss created a simple keyboard that dolphins could use to request objects (fish, ball, ring) or an activity (rub). Although the two adult females in the pool, Terry and Circe, essentially ignored the keyboard, their calves, Pan and Delphi, pushed keys. One indication that they understood the relationship between the key pushes and their outcomes was their preference for the fish key, a key that had to be discontinued due to its overuse to the exclusion of other keys. Although the keyboard did offer the dolphins some control over their environment, its role as a communication device was limited. The most interesting outcome was an analysis of the calves' whistles in the keyboard context conducted by Reiss and her associate Brenda McCowan. They found that sometimes the calves imitated the keyboard-generated whistles, or parts of the whistles, associated with the keyboard objects when they were playing with those objects.

In the 1990s John Gory, Mark Xitco, and Stan Kuczaj worked on a project to create a large keyboard designed specifically for two-way communication. Two dolphins, Bob and Toby, tagged along with their SCUBA-equipped trainers as the trainers broke infrared beams in the large keys (fashioned with echolocatable objects like shells and metal bowls to make the keys easy to discriminate) leading to a woman's voice saying the English word for each key push. Trainers produced phrases like "herring at shipwreck," which meant that there were herring in some receptacle on the floor of their multispecies habitat in Orlando, Florida, where there were remnants of items associated with a shipwreck. After about six months of swimming with the trainers to the keyboard and then off to different places in the habitat such as the shipwreck, divider, restaurant, etc., the dolphins began to use the keys themselves, and the trainers responded as if the dolphins understood what they were saying with their key presses.

This method of "rich interpretation" of the dolphins' utterances is associated with children's language acquisition: children grow up in a complex environment in which language is modeled, and the models support the children's learning in part by finding a way to respond to their utterances as though there is clear communication. In this case, if the dolphins activated the "herring" key, the trainers responded by keying "herring at restaurant" and then swimming off to the restaurant (a window connecting the underwater habitat with a restaurant).

DOLPHIN INTERACTION

Although it's still not clear exactly how much Bob and Toby knew about the keyboard (videos are still being analyzed), a very cool dolphin communicative device did emerge from this project. The dolphins started to use their rostra (their "bottlenoses"—mouths and jaws) to point out items of interest to the trainers. The items of interest were usually—surprise!—jars in which there were fish (for example, herring and smelt), and the trainers engaged in this way would generally unscrew the jars and let the dolphins have the fish inside them. Perhaps you are wondering whether their assertion that the dolphins were "pointing to communicate" with their trainers is too much of a "rich interpretation" of their behavior? If so, good for you! It is very easy to overinterpret animal behaviors by putting oneself into the animal's shoes, and anthropomorphizing their behaviors, that is putting them into human terms (for example, "shoes"?!?). In this case, however, the evidence is pretty strong that Bob and Toby were indeed trying to communicate. For one thing, in 461 of the 722 points, not only did Bob or Toby stop swimming and point to the object but they also turned their heads back toward the trainer and then back to the object again. This monitoring suggests that the dolphins hadn't just learned that "if I align my body with the object, I get a fish" but also that the trainer had to notice. Another piece of evidence suggesting that the dolphins were communicating with people was that they never pointed when humans were not in the water even though the rest of the setup was the same. In a follow-up confirmation study with baited and unbaited jars and trainers who faced the dolphins, turned their backs, or swam behind a barrier, the dolphins were significantly more likely to point when the trainer was face-forward and could see them and respond to a point—again suggesting that the dolphins needed a receptive receiver before initiating a point.

Below A dolphin observes a trainer activating a key on a dolphin-friendly keyboard. Here dolphins learn to use the keyboard by watching their human companions via a project modeled on interspecies communication between primate species as depicted on page 51.

In contrast to the previous projects in which humans interacted with dolphins in human habitats, Denise Herzing and her colleagues introduced a keyboard to wild Atlantic spotted dolphins (versus the bottlenose dolphins who are the focus of the rest of this chapter) who had spontaneously interacted with humans in Herzing's long-term study of these dolphins around Little Bahama Bank in the Caribbean. Visual keys and computer-generated tones were associated with a scarf, rope, sargassum, and a bow ride. Although humans interacting with each other at the keyboard appeared to draw some dolphins into engagement in the activities surrounding it, the impact of keyboard use specifically on the dolphins was hard to gauge and resulted in work by Herzing to create a more acoustically based system which she is currently trying to implement.

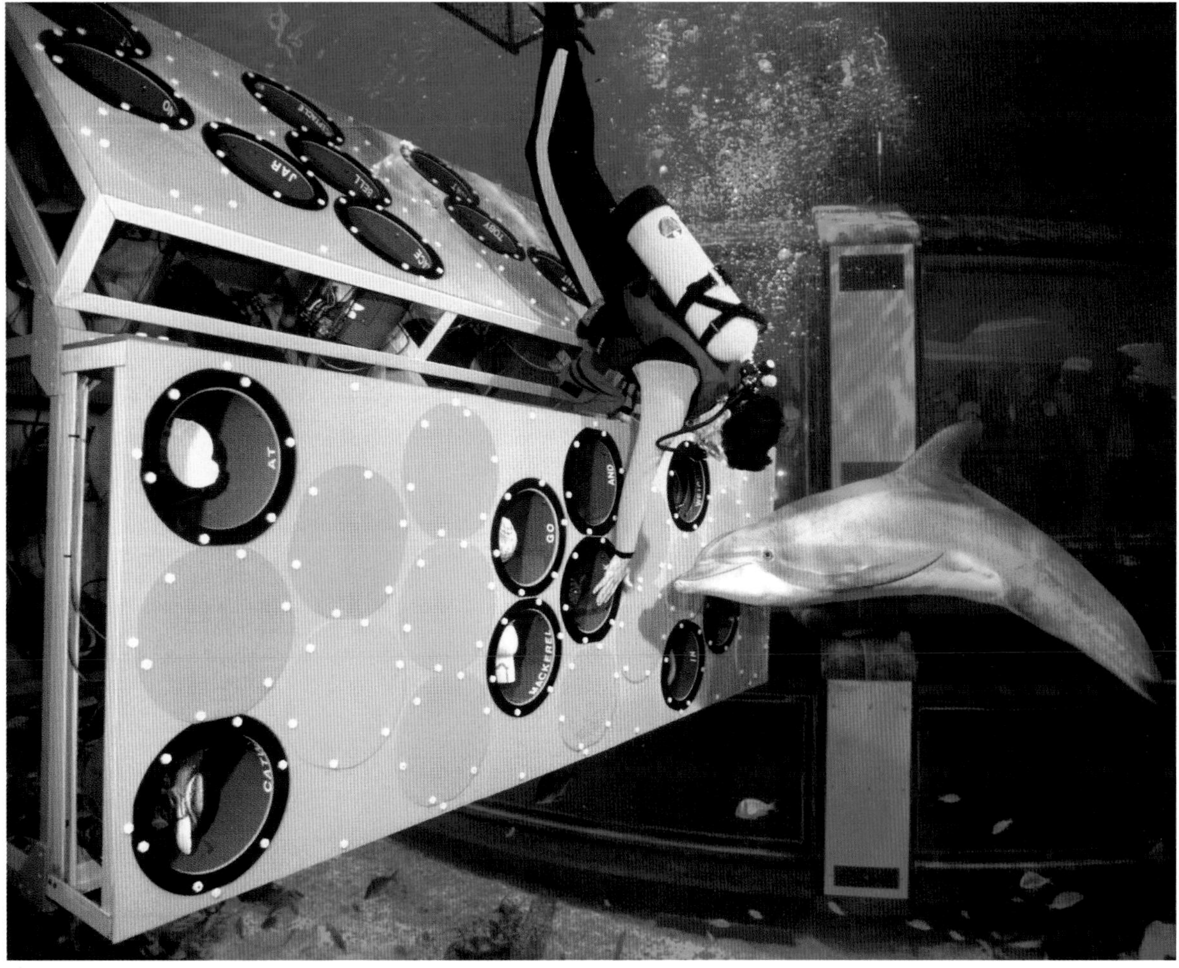

LEARNING TO THINK LIKE CETACEANS

Although the communication studies with dolphins outlined on the previous pages have uncovered some of their communicative skills (for example, pointing when humans can see the dolphins versus when the humans cannot, responding differentially to object and action designators when they change positions in a sequence, and spontaneously imitating some object labels), there are still many unanswered questions about their capacities in this arena. In most cases these projects engaged questions related not only to what the dolphins could do but also what their responses indicated about their capacities in relation to human language—big philosophical questions concerning reference, symbolism, and how language itself might affect humans (that is, questions that humans have been asking for millennia and that are, not surprisingly given their scope, still under active debate). So far, our fairly simplistic human-dolphin communication projects have not moved the mark too compellingly in these queries compared to studies of language use across human cultures (for example, the Amazonian Pirahã's use of language is a very interesting counterpoint to Western cultures with the Pirahã's focus on direct experience and lack of references to the distant past and future) and communication projects between humans and nonhuman primates. (If you're reading this book and you haven't read about Sue Savage-Rumbaugh's work with the amazing bonobo Kanzi who not only produces phrases with "lexigram" shapes depicted on a board and on computer keys but also understands both lexigram phrases and English ones, a happy opportunity awaits!) You may wonder why we have done so much less with dolphins than we've done with nonhuman primates, but here your question likely reflects the answer. Because humans are primates, it is easier to work with other primates. We primates tend to live on the land, have fairly similar sensory systems, manipulate objects with our hands, make facial expressions, eat a wide range of shared foods, sleep with both hemispheres

of our brains, and can generally share a great deal of the same technology—headphones, joysticks, buttons, buildings, computer screens, forest paths, and on and on. Dolphins, on the other hand, live in the water, exponentially outclass us in sound processing (but have no olfactory bulbs for processing smell), have no hands and little in the way of facial expressions, eat a limited variety of sea creatures that they mostly swallow whole, and appear to sleep one brain hemisphere at a time. They also require very sophisticated equipment for study (for example, sound equipment that receives and produces sound much higher in pitch and faster than the equipment built for humans, boats, SCUBA, and rustproof objects), and expensive interfaces for interaction. And as hard as all of that is to manage, the hardest issue is taking a cetacean-centric approach to studying them versus an anthropocentric one. Which questions are the best ones to ask of dolphins? Are ours too primate-centric? The next section tackles some of these issues across a range of approaches.

Opposite A bottlenose dolphin engages in a fish-matching task using vision alone. The dolphin sees a cut-out of a fish on the left, and he is then rewarded for choosing the matching species presented in videos on the right. Some matches are easier than others. Scientists can use the dolphin's confusions to better understand which features of the fish dolphins are using to tell them apart.

Below Primatologist Sue Savage-Rumbaugh and a bonobo use a keyboard to communicate with each other. Dr. Savage-Rumbaugh established that these apes, like children, learn by watching and that, without specific training, they too could acquire the use of symbols by observing others using those symbols in day-to-day communications. Through her lifelong work with primates Dr. Savage-Rumbaugh has forced us to reconsider which traits are truly and distinctly human.

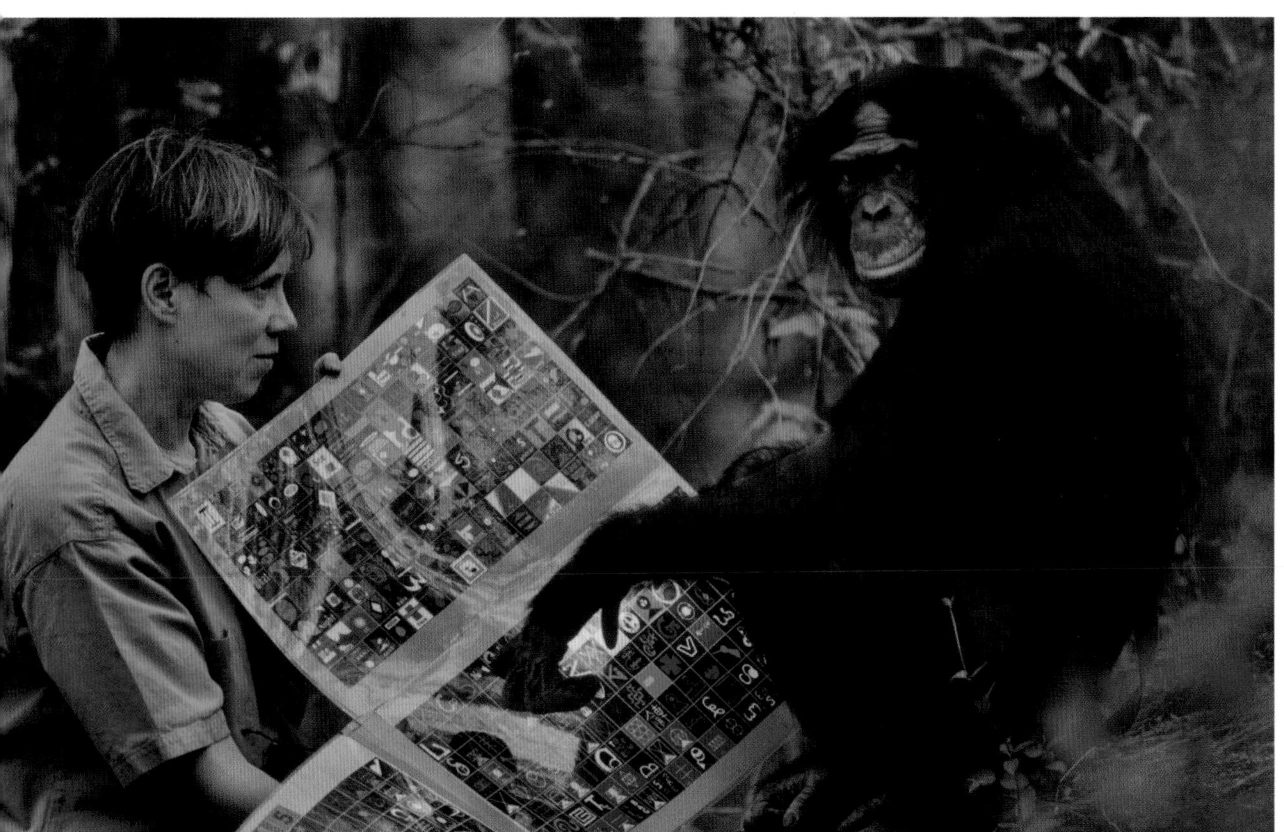

HUMANS KNOWING DOLPHINS

The communication systems humans created to learn about dolphins likely taught the dolphins as much, or more, about us as we learned about them. Most of the systems, although often originally envisioned as two-way, ended up being all, or much more, we-to-them than they-to-us. Reasons for this asymmetry may be that, first, we humans made the choices and so we knew our goals and the dolphins may or may not have understood them or been motivated by them and, second, we humans tried to incorporate sound into the systems, but we tended to be poor at tracking any sounds they returned to us in real time. In this section, we'll focus on how scientists have tried to determine more about how dolphins themselves experience the world through sound, clearly an important avenue of information for them.

Maybe the best way to begin is through your own imagination and memory because to become more dolphin-centric we need to begin with the sea. Take a moment here to imagine or remember being in the ocean. At the surface you may notice the glitter of the light on the chop or the cries of the sea birds. As you dive down a couple of yards and open your eyes, what do you notice? Perhaps first it's that you almost have the ability to fly. It's easy to move from deep to shallow; gravity doesn't keep you down like it does in air. Perhaps next it's that the light is dimming and that sound is coming from everywhere. Maybe at this point you are beginning to feel a little panicky. Why? You need to breathe! But you've been flying and you're a little disoriented. Yikes! Where is the surface?!?

Finding the surface is important when you're an air-breather who spends all your time in the ocean. And while there are a lot of cues to help you get there (pressure changes, light changes, flora and fauna), one easy option for dolphins is the use of echolocation, a form of biosonar in which dolphins send out clicks and receive echoes from their environment. Dolphins produce echolocation clicks within their blowhole apparatus, send them through their rounded heads, and interpret the echoes that return from their surroundings. Air pockets produce loud, clear echoes, and so the surface should be easy to find in just about any circumstance. In general, sound is an excellent energy source on which to depend underwater because it travels so efficiently there—more than four times faster than in air.

A SONIC WORLD

Even for the primate-centric, it's easy to recognize that sound is central to how dolphins negotiate and know their world. Much has been made of the dolphin's large brain (large in absolute terms and in relation to its body size, see Chapter 2) both within and beyond science, and many hypotheses have emerged over the years about how dolphins might use their large brains. These hypotheses have ranged widely in explaining the brain's central functions: from a method of dealing with heat loss in the water to processing echoes in microseconds to handling complex social interactions to being spiritual leaders. All of these suggestions lead one to wonder how the dolphin represents its world: What information is available? What stands out?

Opposite top A dolphin responding to commands from a researcher using an electric pinger working in the Caribbean Sea. Because pingers can produce clear signals that travel well underwater, these devices serve as efficient calling mechanisms between humans and dolphins working together in open-ocean environments.

Opposite bottom Present-day mammals connect across a short distance and a long evolutionary divide.

If one is searching for compelling cetacean-centric cognition questions, those related to echolocation are both slow-moving fish (in primate terms, "low-hanging fruit") and intriguing in that echolocation feels rather foreign to us and it has interesting cultural implications. For example, through the years scientists have speculated about the etiquette of dolphins' using echolocation in a group because it's an active sensory system. Humans also have sensory systems that are more active or more passive. For example, hearing is more passive—you might hear your neighbor's cell phone conversation whether you want to or not. On the other hand, touch is more active—you usually need to reach out and do it, and it's evident to a bystander which object you're investigating through touch. (This is one reason there are so many cultural rules about touching.) Similarly, when a dolphin echolocates, she must actively send out clicks, meaning that a bystander could potentially tell which object she's investigating. Is this what really happens? Mark Xitco and Herb Roitblat checked it out—and you can too by reading the case study testing echoic eavesdropping, on the following pages.

Right A group of common dolphins appears to work collaboratively to herd a shoal of sardines into a bait ball. Individual animals then plough through the compacted ball to feed on fish that try to break loose.

CASE STUDY: **ECHOIC EAVESDROPPING**

When dolphins echolocate, they produce loud, short clicks that are directed in a very focused beam in front of their heads. This eruption of sound could be rather obvious to anyone swimming nearby, and so Mark Xitco and Herb Roitblat wanted to find out if a non-echolocating dolphin could get information about an object that a neighboring echolocating dolphin was investigating.

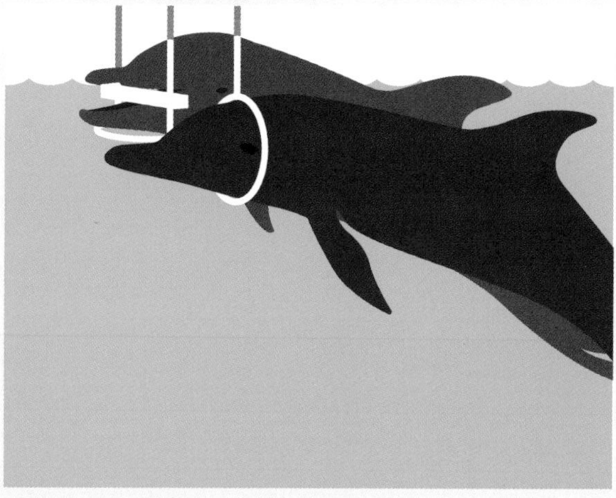

1 ECHOLOCATING & NON-ECHOLOCATING

Two male dolphins, Bob and Toby, adopted separate roles in an echoic matching task. In this task, a sample object hung behind a thin visually opaque but echoically transparent black plastic screen that allowed Bob to echolocate, but not see, the sample. Toby positioned himself right next to Bob such that he was close enough to intercept the echoes coming back from the sample object but could not echolocate it himself because his big melon—the part of his head through which the dolphin's echolocation clicks travel from the blowhole into the water—was in the air but his lower jaw, the part of the dolphin's anatomy that receives echoes, was underwater.

2 MELON ABOVE WATER

Notice that the top of Toby's head is above water whereas Bob's entire head is submerged as they position themselves in hoops in front of the sample object. After Bob echolocated a sample object and Toby "eavesdropped" on the echoes, both dolphins swam off in different directions to their own arrays of the same three objects; each array included a copy of the sample object the dolphins had just accessed through echolocating and listening (Bob) or just listening (Toby).

3 ATTEMPTING TO MATCH THE OBJECT

The results were rather exciting because it turned out that Toby was indeed getting enough information to choose the correct object at above-chance (better than just guessing) accuracy levels when both dolphins were familiar with the objects. This finding taught us that a non-echolocating dolphin can interpret the echoes of a neighboring echolocator. We also learned that dolphins swimming together probably know what their neighbors are paying attention to when they're echolocating. Want to hide the fact that you've found a fish from your fellows? Doesn't look like a realistic option if you're a dolphin. Maybe this is one reason why dolphins hunt cooperatively so well.

4 THE CHOICE OF OBJECTS

Both dolphins received some fish if they chose the object in the array that matched the sample.

After Xitco and Roitblat's study, Tomas Götz and his colleagues recorded the echolocation clicks of wild dolphins swimming in dense and sparse groups and discovered that fewer dolphins echolocated when they were close together than when they were more dispersed—perhaps because they could listen in to each other's echoes and/or perhaps because too many echolocators close to each other would cause too much cacophonous interference among the returning echoes.

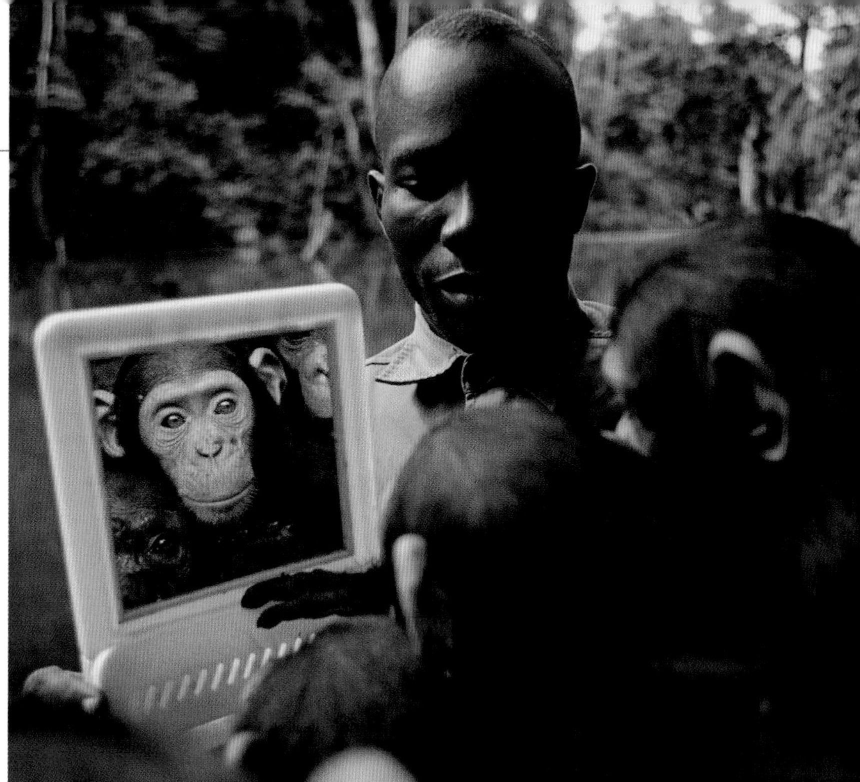

Right A trainer holds up a mirror to a group of chimpanzees in a reserve. Human and nonhuman primates can use vision to connect.

UMWELTEN

Social echolocating is a fascinating focus of study, but the study of independent echolocation is also a worthy pursuit of cognitive research. It is interesting not only because humans would like to be able to be as good at sonar as dolphins are (right now, we're not even close), but also because as a cetacean-centric query, it lets us glimpse the dolphin's world.

About a hundred years ago, a German named von Uexküll highlighted an intriguing idea—that different species, while inhabiting the same planet, actually live in different worlds. No, he wasn't crazy or even a sci-fi writer. Rather, he was acknowledging that because different species need to do different things to survive, they have evolved a variety of sensory systems to help them get those jobs done. In turn, those sensory systems give them privileged information. For example, dogs have a very intricate olfactory system. My guess is that because they have so much more sensory and brain tissue devoted to smell than we do, their experience of smells is beyond our imagination; it's qualitatively different from our own. It would be interesting—although perhaps not always pleasant—to smell the world that a dog smells. As it is, we might share a space with a dog, but we're missing a lot of what she's experiencing. Dogs and humans are living in different worlds, different umwelten to use von Uexküll's term.

As marine echolocators, dolphins also live in a very different umwelt from ours. So if we want to know what it's like to be a dolphin and to know more about the uses to which they're putting all that brain power, we need to study what information they're getting about the world through echolocation.

The case study on page 60 explores the dolphin's umwelt by investigating the information the dolphin gets about objects through vision and echolocation. One function of a sophisticated brain is the representation of detailed information. In this experiment we learned more about how and what dolphins represent about their world, and we did it using vision as well as echolocation in part because it makes the dolphin's representation more accessible to humans since we're so good at getting information visually. On the other hand, sometimes our understanding of echolocation is too visually oriented; we talk about echolocation as seeing with sound. That's to some extent true for humans; we create pictures from ultrasound in order to see a fetus in its mother's womb, for example, and we can get quite detailed information through this technology such as the number of chambers in a baby's heart. However, that's quite a different system from a dolphin's because the ultrasound machine's "clicks" are higher in pitch and occur much more frequently in time compared to a dolphin's clicks. The machine also has an enormous memory in which to store information from echoes as points in a picture. Dolphin echolocation

Left An Atlantic spotted dolphin carrying seaweed, possibly to impress females. Using objects for sexual display is rare among animals other than humans and chimpanzees. Dolphins more often rely on sound to negotiate their world.

likely does not work like that since it's constrained by biology. The dolphin can't produce millions of clicks per second, the neurons don't fire as fast as electronics, and short-term memory isn't as big in an animal as it is in a computer. Nevertheless, real animals can do it all—move, get information in the world through sensory systems and turn it into neural information, think, and respond in surprising ways. Go life!

Now, about that dolphin experiment. From the case study, you'll see what the dolphin did. The design of the experiment is a little funny in that we investigated how echolocation works by finding out if we could lie to the dolphin. Seems like a questionable strategy? Actually, it worked! Here's why. A classic way of doing science is to pit two theories against each other and find out through experimental design which theory the data support. In this case the two theories are about sensory systems: one, the dolphin learns to associate its visual and echoic experiences in the same way that humans learn to associate, for example, the smell of a flower and the way that it looks; or, two, the dolphin recognizes that it's getting information about the same object feature (for example, size) through both vision and echolocation.

ASSOCIATION & RECOGNITION

How does lying fit into the experimental design? Well, if I ask you to close your eyes and smell a flower you don't know—let's call it a roseydaffodaisy—and then I show you two flowers, the roseydaffodaisy along with a pricklytulilily, and I lie to you and tell you that you just smelled the pricklytulilily, you'd believe me, right? Because there's nothing about how the flower looks that tells you about how it smells. You learn to associate the look of the flower with how it smells.

On the other hand, if I ask you to close your eyes and feel a weird wheel-like object that's smooth on one side and rough on the other—let's call it a prongpie—and then show you the prongpie along with a mudpie alien covered in spikes and marshmallows and I tell you that you'd felt the alien, you wouldn't believe me. You'd reply, "No, I felt a wheel-like object that was smooth and rough, not a spiky-marshmallow-mudpie alien." (Clearly, you are very good at describing objects specifically!) I couldn't successfully lie to you in this case because you can recognize an equivalence between the object attributes you feel and those you see. You can recognize smooth and rough, spikes and puffs, both when you feel them and when you see them. With the dolphin we used a similar design to find out if dolphin echolocation is more like vision and smell in humans (learned by association, theory #1) or more like vision and touch in humans (recognition of equivalence, theory #2). You can see the way we conducted the experiment from the case study on the following pages.

CASE STUDY: **ECHOLOCATION & VISION**

Dolphins live in a sound-dominated world, but, of course, their other senses are also important to them. Their sixth sense, echolocation, remains somewhat mysterious to us; we do not know how they decode an object's echoes to get attribute information about that object. To understand more about what it's like to be a dolphin, we can enter their echoic world by studying the information they get through echolocation and also through vision, a sensory system more familiar to humans.

1 DOLPHIN VIEWING SAMPLE OBJECT
Echolocation isn't really picture-building, rather it's a way of listening to get information about an object. To find out what kinds of information dolphins get echoically that could also be accessible visually, the dolphin Toby saw an unfamiliar sample object in air where he couldn't echolocate it because the density difference between water and air precludes dolphin echolocation.

2 ECHOLOCATING ALTERNATIVE OBJECTS
Next Toby echolocated but could not see a set of three objects that included the sample he'd just seen, positioned in a row behind thin black plastic through which echolocation clicks and echoes could travel. We also reversed this scenario so that with some objects Toby echolocated the sample and then saw the alternatives.

Clockwise from top left Horn, cup, baby bottle, pair of foam cones, elephant, helmet, plunger, tree-shaped baking pan, and wood massager.

3 MATCHING UNFAMILIAR OBJECTS

Across the experiment there were six 9-object sets unfamiliar to Toby. Usually Toby received a fish for matching the same objects, X to X. However, within each large set there was one pair of objects for which he received a fish for matching two different objects, Y to Z and Z to Y. This design allowed us to see if we could lie to Toby about the relationship between what he was seeing and what he was echolocating. We were pitting two theories against each other: Does the dolphin learn to associate his visual and echoic experiences like humans learn to associate, for example, the smell of a flower and the way that it looks; or recognize that he's getting information about the same object feature (for example, texture) through both vision and echolocation. You can learn more about how this experimental design helps us understand the dolphin's world by reading pages 58–9.

4 A FISH IS A FISH IS A FISH

We discovered we couldn't lie to Toby. He chose matching objects even without a fish reward 498 out of 672 (74 percent) times (even though we reinforced Y to Z, he chose Y after Y and Z after Z). We couldn't fool him. For dolphins, echolocation and vision don't work like human smell and vision but rather like human touch and vision with overlapping information. Dolphins recognize an equivalence between what they see and echolocate. This is good news for humans who want to understand more about dolphins: our sensory worlds are similar in some ways, and one function of the dolphin brain is to represent the world in its glorious multifaceted complexity as we do.

KNOWING WHAT DOLPHINS KNOW

Mind, the mysterious frontier. These are the voyages of the Science-ship *Discovery*. Its continuing mission to explore strange new worlds, to seek out new perspectives and new ways of thinking, to boldly go where no one has gone before...

As I write these final paragraphs of this chapter, I wonder who will be reading it and if it took you where you expected to go. Do you recognize the basis of the *Star Trek* quote I rewrote at the beginning of this section? I predict that you do recognize the *Star Trek* quote and that you did not expect what you read in this chapter. Why? Because when it comes to dolphins, popular science and science fiction have blended to create a picture of dolphins that clouds their real nature. Although there is a place for just imagining what a dolphin experiences—in the scientist's imagination, in an artist's imagination, in a writer's imagination, in your imagination—science gives us the tools to test the hypotheses that might develop from our conjurings in order to find out how close to the dolphin's experiences they might be.

Sometimes I wonder if it wouldn't be more fun to imagine the dolphin's mind than to study the real thing, but for me it's turned out to be much more fun to enter the dolphin's world by asking the dolphin.

This chapter gives merely a taste of what we know about what dolphins know, and what we know is only the barest hint of what there is to know. Through two-way communication systems, dolphins have learned about human systems and humans have learned about the dolphin's capacity to get information from sequencing, to remember strings of signs and sounds, and to point in order to direct humans. Through matching studies, humans have learned that dolphins can use echolocation to track each other's attention and acquire the detailed information needed to create an intricate sensory world. The current zeitgeist suggests that studies of human-dolphin partnerships may be a major focus for some near-future work. Why have both species entertained each other's minds for millennia? Perhaps we'll find out.

Left Inquisitive, highly intelligent, and sociable, common or Atlantic bottlenose dolphins are probably the most easily recognized of all marine mammals. They commonly form groups of between 2 and 15 individuals, feeding singly or cooperatively in shallow coastal waters, although offshore they may hunt in groups numbering several hundred individuals.

Opposite Spinner dolphins are a highly social species and often hunt cooperatively in groups at night and rest in shallower bays during the day. They are capable of diving down to almost 1,000 feet (300 meters) and can remain underwater for 5-10 minutes before surfacing to breathe but to minimize their diving efforts they tend to hunt at night when their preferred prey of fish, squid, and shrimp move up from the depths.

IS THAT DOLPHIN IN THE MIRROR ME?

Cogito ergo sum: I think, therefore I am.

When the seventeenth-century philosopher René Descartes doubted his own existence, this statement helped to reassure him. How could he doubt, a form of thinking, if he did not exist? It's not surprising that Descartes began his intellectual journey with an acknowledgment of himself as thinker. Part of being human includes a strong sense of self-identity. A great deal of our lives is occupied with our sense of ourselves—our gender, ethnicity, occupation, and social roles as seen through names, languages, religions, nationalities, clothes, transportation, food, etc. Self-concept is so central to humans that we often assume that other species represent themselves as we represent ourselves. "Look at Fido. He knows he looks sharp in that sweater!" or "Awwww, look at Fido. He knows he shouldn't have done that!" But as it turns out, Alexandra Horowitz discovered that dogs are apt to put on a "guilty" look when they hear a chiding tone whether or not they have engaged in forbidden behaviors. Why? Apparently because their owners are more likely to go easy on them when they appear to humans to be ashamed, according to some work by Julie Hecht. So puppy-dog-eyed dogs don't necessarily represent their guilty selves at all. Rather, they are guided by the outcomes—reinforcement and punishment—of their stances as interpreted by humans.

And dolphins? Do they represent themselves? The studies most focused on this question have generally used a "mirror-mark" test to investigate it. The mirror-mark test was initially developed by Gordon Gallup in 1970 to determine whether a chimpanzee could recognize himself. First Gallup watched his chimpanzees and categorized their behaviors as either social (directed toward others—like vocalizing) or self-directed (directed to themselves—like picking food from between their teeth). Then the chimpanzees were anesthetized and marked with a dye on their eyebrow ridges and ears, areas that were only visible to a marked individual when looking at a mirror. Once the chimpanzees awoke, Gallup counted how often each chimp touched the dyed areas on him or herself when a mirror was present or absent. Gallup's chimpanzees touched the marks more often when the mirror was present, although chimpanzees tested by other researchers in similar setups did not always engage in the same way. Nevertheless, the test is common for questions about self-recognition in young children and animals, and so a couple of researchers, Diana Reiss and Lori Marino, worked with Gallup—and later without him—to test dolphins using what they hoped was an analog of this procedure. Of course, the dolphins could neither be anesthetized (they're voluntary breathers) nor touch themselves (the pectoral fins on their sides are pretty short!), so Reiss and Marino tried something new. They marked the dolphins with a hue in some trials and with a clear "sham" mark in others with the idea that the dolphins would spend more time in front of the mirror and engaging in self-referential behaviors when the marks had a hue versus when they did not. As it turned out, the results of the study were difficult to interpret because the dolphins spent a fairly small amount of time in front of the mirror in general but a good deal of that time was when they had been marked with the clear shams. Across several mirror studies, it was also difficult to determine when behaviors were social versus self-directed. Gallup himself suggested that this method was not a good fit for dolphins, and in truth their sleek physique, acoustic orientation, and visual system (their eyes are on the sides of their head) make this method a hard sell. In addition, the dolphins' motivations are hard to decipher because even when they do spend time looking at themselves in the mirror and responding to the mark, they can be doing it because they enjoy controlling a visual stimulus—the mark speeds around as they move—rather than because they know that mark is on their own bodies. Do dolphins have a self-concept? At the moment, only dolphins would know for certain.

4 CETACEAN COMMUNICATION

Laela Sayigh & Vincent Janik

Untitled, Humpback Whale Calf
Kingdom of Tonga, 2005

"After numerous days observing this mother and calf pair, my most memorable moment was when the calf rested its body against the sandy bottom seven meters below. The mother glided over the calf and gently rested her body on top of him. She released a small amount of air, where they then remained on the bottom together for over four minutes."—Bryant Austin

HOW DO CETACEANS COMMUNICATE?

The idea that cetaceans might have "language" similar to our own has long been a source of fascination to scientists and lay people alike. The person primarily responsible for bringing the idea of language in dolphins into the public eye was the scientist John Lilly (see pages 44–5), who in the 1960s and 1970s wrote several popular books that made sensational claims about whale and dolphin communication and intelligence. But he could not back these up with any real scientific evidence, and the interest of science funders dwindled. So what do we really know about how these animals communicate with each other? Less than you might think!

Cetaceans communicate with each other primarily by using sounds. There are few other ways to send signals underwater, since visibility is usually poor and smell cannot be used when holding your breath. Taste and touch might work, but these only allow for relatively simple messages. Sound, on the other hand, travels very efficiently, and some species use frequencies that can travel over long distances (tens or even hundreds of miles), making sound the best avenue for communicating underwater. As we shall see, cetaceans use sounds in every aspect of their lives: for mating, feeding, and keeping in contact with family and friends.

While sound is easy to record, it is not that easy to study its role in communication. To tell who is calling and who is responding, communication scientists often observe which animal opens their mouth or beak. But cetaceans do not need to open their mouths when they call—sound can just travel from its source inside the animal through the skin to the outside world. Therefore, more sophisticated methods are needed to localize who is calling. Another problem is the large range of acoustic signals used by cetaceans. If caller and receiver are miles apart, it is hard to observe both at the same time. Despite these challenges, scientists have gained key insights into cetacean communication by using modern technology, especially for some of the more accessible species.

MAKING SOUNDS UNDERWATER

Have you ever tried speaking underwater? For a terrestrial mammal, it is not so easy to produce sounds when submerged. We mostly produce a stream of air bubbles if we try. Some of the sound is audible underwater, but much of it stays in the air bubbles. This happens because sound is reflected at the boundary of two media that have very different densities (like air and water—molecules in water are much closer together than they are in air, so water has the higher density of the two).

So, how do cetaceans get around this problem? They still have the mammalian voice box, the larynx, but it is ill-suited for making sound underwater. The biggest challenge is to get the water (rather than the air that terrestrial animals use for sound production) to vibrate. The solution, at least for odontocetes, is a new structure to produce sounds. Odontocetes have nasal passages in their foreheads that allow them to breathe out of the top of their heads rather than having to stick their whole face out of the water to breathe like we do. These nasal passages are very special because they are not only used for breathing but also for producing sound. They use two pairs of liplike structures and large air cavities in their nasal passages to recycle air by pumping it back and forth between air reservoirs. Imagine connecting two balloons and moving air back and forth from one to the other and you get the picture. The liplike structures used to be called monkey lips, because they really do look like those of a primate. Now they are called phonic lips. By forcing air through these closed lips, odontocetes can produce clicks and whistles. When doing this, they produce vibrations in two media: the air that they use, and more importantly, the lips themselves. The lips are made of tissue, which is comprised largely of water. By producing sound this way, acoustic energy in the tissue can travel through the body of the animal and into the

Below Humpback mother, calf (right), and escort. Male humpbacks are notable for the complex songs that they produce on the breeding grounds.

surrounding water without being reflected. To optimize this sound transmission through the body, odontocetes have special "acoustic fats" in a structure called the melon in their foreheads. These acoustic fats are even more like water in their density than other tissues and can direct sound efficiently into the environment.

HEARING SOUND

The same problem of reflections between water and air occurs on the receiving side. To experience this yourself, try diving down in your bath while someone is talking to you in the room. You won't be able to hear much once you are submerged because most of the sound in the room is reflected off the surface rather than entering your bath water. The problem here is that our ear is adapted to hearing in air. However, hearing sound in water is a little easier than producing it, because the mammalian inner ear, where we really perceive sound, is already filled with a fluid much like water. The human middle ear is a special adaptation to maximize the transfer of sound energy from air to the fluid in the inner ear. Whales and dolphins don't have to make that transfer. But they still need to direct the sound to the inner ear, something that the human outer ear is designed to do. In odontocetes, there are special areas in the lower jaw that are filled with more acoustic fats that receive and direct sounds straight to the oval window that leads into the inner ear. However, sound can still travel through all other parts of their bodies, which makes it hard to tell where sound is coming from. To avoid this problem, dolphins and whales have an enlarged bone structure around each ear called the bulla. The bulla exists in other mammals as well, but it is usually much smaller. The bulla allows sounds from the fatty acid channels in the jaw to reach the inner ear, but it shades the ear from sounds coming from other directions, enabling cetaceans to tell which direction a sound is coming from. The large bulla around the ears of cetaceans is so unique that it is seen as one of the defining features for the whole cetacean order.

We know much less about how mysticetes make and hear sounds. They have additional air cavities to recycle air, but they lack the monkey lips used by odontocetes. We still need to locate the actual vibrator for mysticete sound production. Given that they usually produce much lower-frequency sounds than odontocetes, it is likely to be a larger structure or membrane than the monkey lips. Hearing in mysticetes is likely to be similar to odontocetes, but we still need better evidence to say for sure. It is much harder to study hearing or sound production in an animal the size of a bus.

Below Newborn gray whale surfacing. Mysticetes are able to make a variety of sounds but the actual mechanism is still not known.

Sound production & perception in dolphins

Dolphins produce both whistles and clicks in the upper nasal passages. Whistles are used in communication and clicks can be used in both communication and echolocation. Echolocation, often called sonar, is used to navigate and to find prey.

The so-called dorsal bursae, of which the phonic lips form part, produce sounds that travel through sound-conducting tissue—"acoustic fats"—in the melon. Sounds are received through acoustic fats in the lower jaw and transmitted to the inner ear. Shown is the auditory bulla, the bony structure that channels sound to the inner ear.

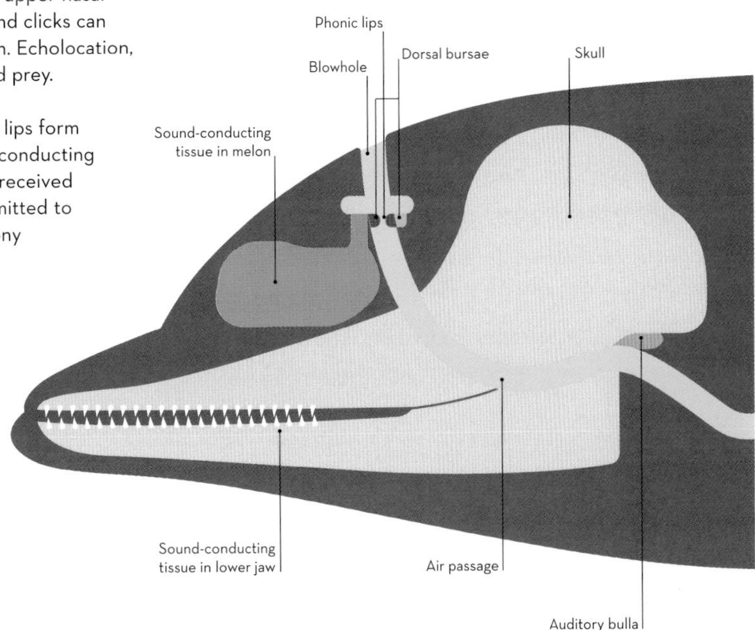

Below Sperm whales, the largest of the odontocetes, produce clicks for sonar and communication, as do dolphins. Their clicking sequences may identify group members and may allow them to coordinate foraging activities.

WHAT TYPES OF SOUNDS DO CETACEANS USE?

With such unusual sound production systems, cetaceans make sounds that are different from most other animals. Their eerie songs and calls were described by early sailors who could hear them through the hulls of their wooden ships, leading to myths of magical sea creatures calling to them. To this day, the sounds of whales and dolphins have a special place in human popular culture. Although underwater recordings can often contain a cacophony of sounds, scientific research has shed light on which sounds are produced by which species. We present what is known about the types of sounds produced by some of the better-studied species of both mysticete and odontocete cetaceans.

MYSTICETES

The song of the humpback whale was among the first cetacean communication signals to be studied intensively, with the first published description in 1971. These studies brought whale song into the public eye, with recordings becoming commercially available and used by a variety of artists in their own compositions. Humpback whale sounds are in the range of human hearing, unlike those of some other species, so they were accessible to researchers using equipment designed for human speech or music. The earliest work focused on descriptions of song structure, which consists of units, which combine to form phrases, which combine to form themes, in a repeating arrangement that can last up to 35 minutes. This repeating structure is what is known as song. Song is only produced by males and likely functions either to attract females and/or to keep other males away.

An intriguing aspect of humpback whale song is that it changes over time, and all males in a given breeding population match these changes and sing the same song. In the northern hemisphere, these changes appear to be gradual, resulting in slow modifications of song structure over time—it can take approximately 15 years for the song to change entirely. However, a faster pace of change was found among humpback whales in Australia, where new song types introduced by western Australian whales into the eastern Australian population were adopted by the vast majority of eastern Australian males within two years. In fact, the same song types traveled from Australia all the way to East Polynesia over several years, without the animals necessarily physically moving that way (see Chapter 6). When whales learn their songs from their neighbors, they seem to prefer to learn from animals that have come from the west. Why this is so is not known. The only mechanism by which song patterns can change so rapidly is by vocal learning, in which whales copy songs of each other. This makes humpbacks one of the few mammalian species to have this skill. The way that humpbacks use vocal learning is very similar to its use in songbirds, another group of animals that is famous for its vocal learning skills (see page 86). However, the mechanism for implementing these changes is not known; for example, which whales initiate the changes, and why they are adopted by other whales.

Calls of several other mysticete species show characteristic temporal patterning that has been described as song, although none have been studied as extensively as humpbacks. These include blue, fin, minke, and bowhead whales. Mysticetes also produce a great diversity of non-song call types, although little is known about their functions. The North Atlantic right whale is one of the better-studied mysticetes species, due to its highly endangered status. The detection of their "up-calls" has become the basis for a ship alert system designed to avoid collisions between ships and whales in the area near Boston, Massachusetts. Whales of all ages and both sexes produce up-calls as well as a variety of other tonal calls. Another call type that has been described for both Southern and North Atlantic right whales is the broad band, impulsive gunshot sound, which is produced by mature males.

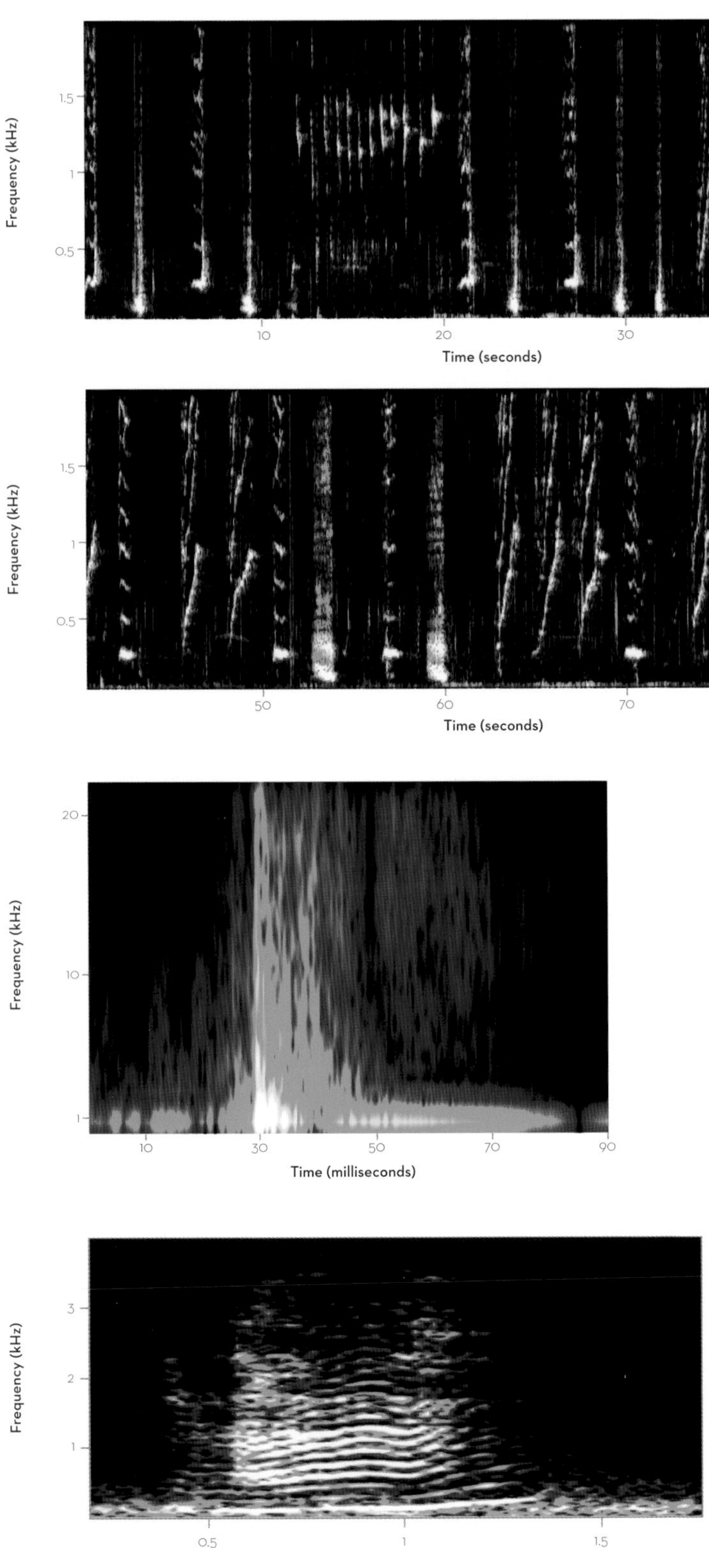

Top and above Spectrogram (plot of frequency on the y axis and time on the x axis, similar to musical notation) of a section of humpback whale song. The two images are continuous, with a total duration of approximately 90 seconds. Note the individual notes or units, which are arranged into phrases. Phrases are repeated, forming themes. Multiple themes compose a single song.

Left and below Spectrograms of "gunshot" (left) and "up-call" (below) sounds produced by North Atlantic right whales. The gunshot is an impulsive sound, with many frequencies represented at one time. The impulse is followed by reverberation. The up-call is a tonal sound that consists of a gradually increasing frequency upsweep, with many associated harmonics (multiples of the fundamental frequency).

How dolphins echolocate

Dolphins use echolocation to locate prey with pinpoint accuracy. The clicks produced in a dolphin's nasal passages are used to determine an object's size, shape, and distance. The sounds bounce off the object and the returning echoes are picked up through sound-conducting tissue in the lower jaw, and transmitted to the inner ear.

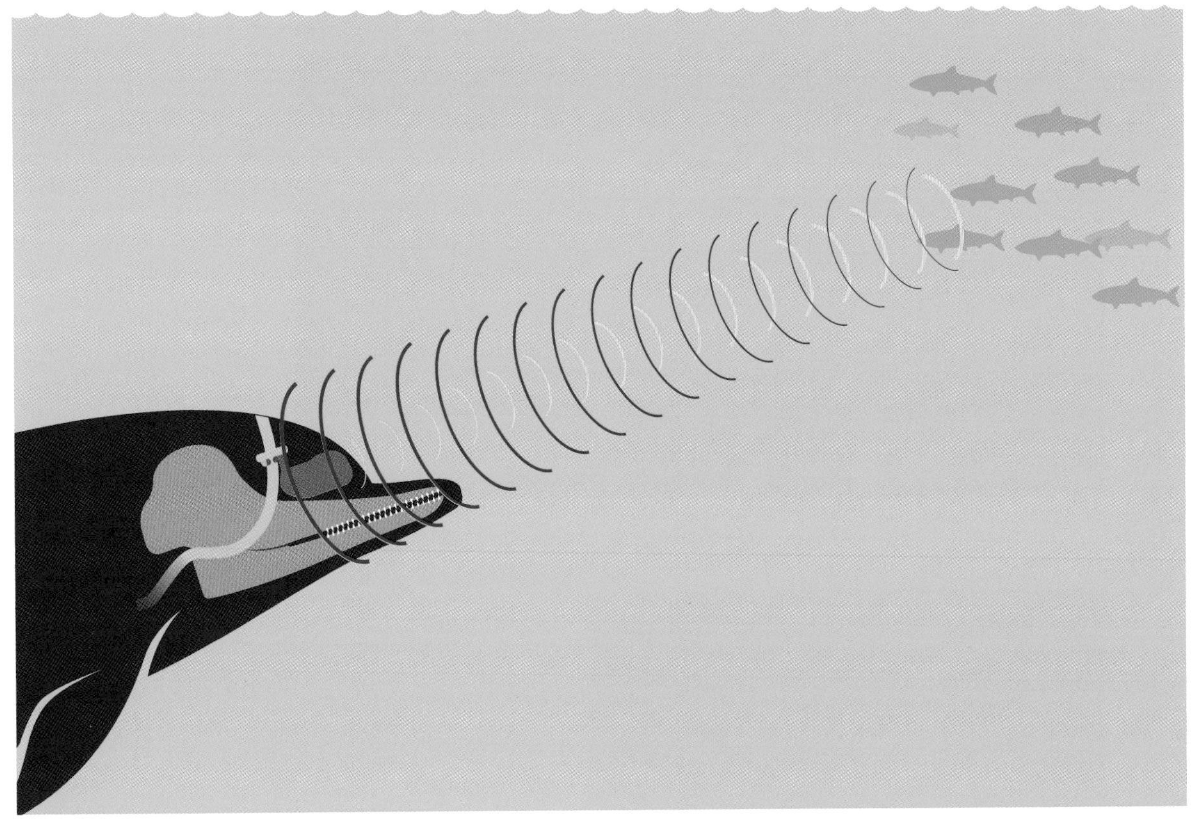

ODONTOCETES

The sounds of odontocetes are typically grouped into three categories: clicks used for echolocation, tonal whistles, and a catch-all category called "burst-pulsed" sounds. Collecting data on the communication of dolphins and other odontocetes is in general more feasible than for mysticetes, because they are smaller and can be briefly captured or studied in captivity. However, although the overall smaller sizes of odontocetes and the higher-frequency ranges of their vocalizations reduce the scale over which their communication likely takes place, these factors introduce new challenges into their study. One of the methods for identifying a calling individual is to use an array of underwater microphones, known as hydrophones, and to use the differences in times at which sounds reach each hydrophone to calculate the location of the individual producing a sound. Another is to use acoustic recording tags that are affixed (noninvasively) to the back of an animal. However, both of these methods are hindered by animals staying in close proximity to group members, since that makes it difficult to tease apart which individual is making a given sound.

ECHOLOCATION OR COMMUNICATION?

All species of odontocetes in which sound production has been studied to date have been found to echolocate, which involves generating acoustic pulses, or clicks, and using information from the echoes of these clicks to essentially "see with sound." Echolocation may function in communication in some or all odontocete species, but this is difficult to document without detailed concurrent behavioral observations. The best evidence for echolocation clicks

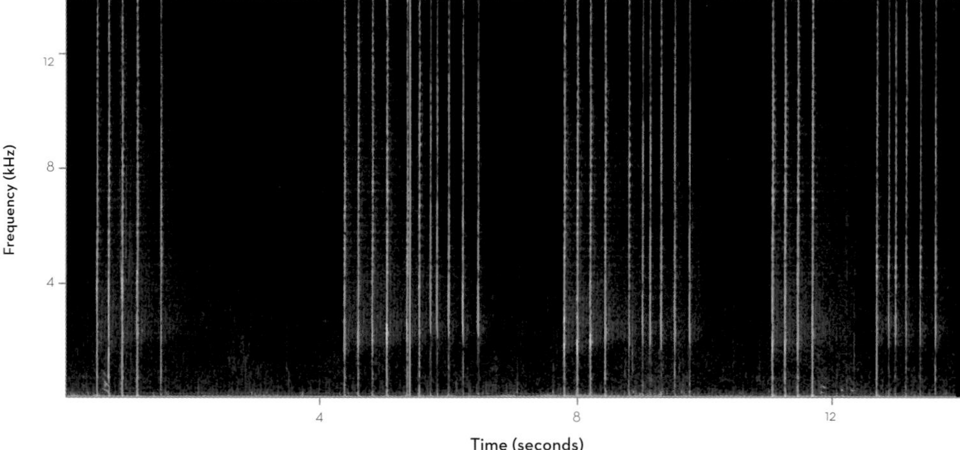

Above Spectrogram of sperm whale codas—rhythmic sequences of relatively low-frequency clicks. This image shows a series of codas produced by more than one whale. Visible are 5-click codas at the beginning and end, and 4-click codas at about 4, 8, and 12 seconds.

serving a communicative function is in the porpoise family, in which several studies found evidence for predictable patterns of click repetition rates in different contexts. These different click patterns may be used to communicate among individuals. Porpoises produce narrow-band high-frequency (NBHF) clicks, which have been described in four diverse odontocete families: Phocoenidae (porpoises), Kogiidae (pygmy and dwarf sperm whales), Pontoporiidae (the Franciscana dolphin, related to river dolphins), and some members of the dolphin family Delphinidae. Species that produce NBHF clicks do not produce whistles, even those in the dolphin family, for which whistle production is otherwise a very important form of communication. Several authors have raised the possibility that NBHF clicks (with an accompanying lack of whistle production) may be an adaptation to reduce predation by killer whales, which rely on acoustic cues to find their prey, but whose hearing sensitivity drops off rapidly at high frequencies above 100 kHz. Similarly, the tendency for some beaked whale species to vocalize only at depths deeper than where killer whales typically occur has also been suggested to be an antipredator adaptation. These vocal adaptations in almost half of the odontocete families suggest that predation may be a very strong selective force in shaping odontocete communication signals.

Sperm whales, the largest odontocetes, produce broad-band, lower-frequency clicks that function in both echolocation and communication. Communicative clicks occur in stereotyped rhythmic patterns called codas. By recording coda repertoires, researchers have identified acoustic clans that are formed by animals that all use the same selection of codas. Animals from different clans seem to avoid each other. Just like in humpback whales, this shared acoustic repertoire is most likely achieved by learning codas from other group members.

Right Bottlenose dolphin in Sarasota, Florida, wearing a noninvasive digital acoustic tag (DTag). DTags are attached with suction cups and stay on for about a day. They record all sounds produced and received by the dolphin, as well as information about its movements and depth.

Above Some of the best-studied odontocetes are members of family Delphinidae. This diverse family includes killer whales (here, a fish-eating killer whale), deep-diving pilot whales, and bottlenose dolphins.

KILLER WHALE COMMUNICATION

Other than sperm whales, the best-studied odontocetes are in the dolphin family, Delphinidae, which contains the bottlenose dolphin of "Flipper" fame, as well as a variety of other smaller and larger species. These include pilot whales and killer whales, which, despite being named "whales," are actually dolphins. Killer whales are top predators in the marine environment, with some preying even on the largest mysticetes. Mammal-eating killer whales have been reported to eavesdrop on the acoustic signals of their prey. They are often very quiet themselves when hunting, to avoid detection by their prey. Fish-eating killer whales, on the other hand, are much more vocal. They produce a variety of pulsed calls and whistles even while hunting. Interestingly, the selection of pulsed calls used by each whale is specific to its pod. Thus, these animals have a kind of dialect in which animals that spend most of their time with each other sound more alike than those that are seen less often together.

BOTTLENOSE DOLPHIN COMMUNICATION

One of the first rigorous scientific studies of bottlenose dolphin communication was carried out by David and Melba Caldwell in the 1960s. While studying captive dolphins, they found that each individual produced an individually distinctive signature whistle that comprised a large portion of its vocal repertoire. Signature whistles have since been the focus of substantial research (see the two case studies in this chapter). Signature whistles have been defined as the most common vocalization produced when individuals are isolated from their group members, but they are also important vocalizations when animals are free swimming. Young bottlenose dolphins learn to develop their own individual signature whistle in the first few months of their lives. Dolphins use these whistles whenever they need to make contact with others. Signature whistles have also been described in several other species such as common dolphins, spotted dolphins, and humpback dolphins. There is even evidence for signature-type calls in non-dolphin species such as beluga whales.

Bottlenose dolphins also produce a variety of non-signature whistles, as well as pulsed sounds (other than echolocation), but few of these vocalizations have been studied in detail. Some of the better-described sounds include the low-frequency, pulsed "pop" vocalization seen in Indian Ocean bottlenose dolphins and "bray" sounds described for bottlenose dolphins in Scotland.

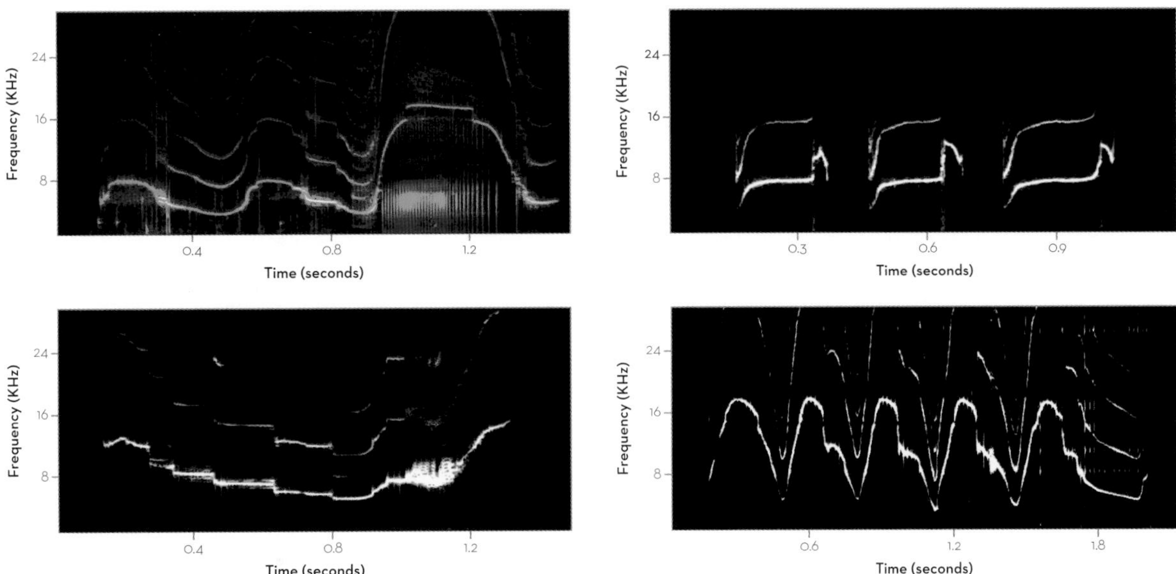

Above Spectrograms of four different bottlenose dolphin signature whistles, showing their distinctive patterns of frequency changes over time. These whistles are typically highly stable over an individual dolphin's lifetime.

Below A group of Indo-Pacific bottlenose dolphins off the island of Reunion. Indo-Pacific bottlenose dolphins are slightly smaller than common bottlenose dolphins and speckle ventrally with age. Like the latter, they are adept at echolocation and produce a range of sounds, including pops and signature whistles.

CASE STUDY: **PLAYBACK EXPERIMENTS**

Since the 1960s, signature whistles have fascinated both the public and the scientific community as they appear similar to a "name." However, a "name" implies that others recognize the signal as referring to an individual dolphin. Thus, a key question is to determine how dolphins use and perceive these whistles. The well-studied resident population of bottlenose dolphins in Sarasota, Florida, provided the ideal opportunity for experiments that involved playing sounds to dolphins and carefully measuring their responses. In all experiments, the responses of dolphins to two different whistle sequences were compared, to account for individual differences among dolphins.

1 INITIAL SET UP

In the first set of experiments, each dolphin heard a sequence of whistles from a relative—often the mother—and a close associate, so that both whistles were similarly familiar. Dolphins consistently responded more strongly—in the form of head turns toward the speaker—when they heard the whistles of their relatives than when they heard whistles of close associates. These results indicate that dolphins use signature whistles to recognize other individual dolphins.

2 WHISTLE RECOGNITION

The next two sets of playback experiments explored the features in whistles that dolphins were using for recognition. Two, not mutually exclusive, possibilities are first, the distinctive pattern of frequency changes over time, or contour, of the whistles, or, second, voice cues caused by the anatomy of the vocal apparatus (often called "byproduct distinctiveness"; this is how you recognize the voice of your mother on the phone without her telling you her name).

NATURAL SIGNATURE WHISTLE

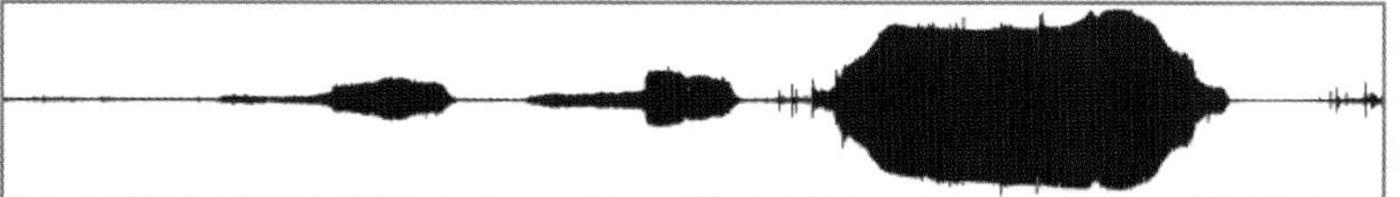

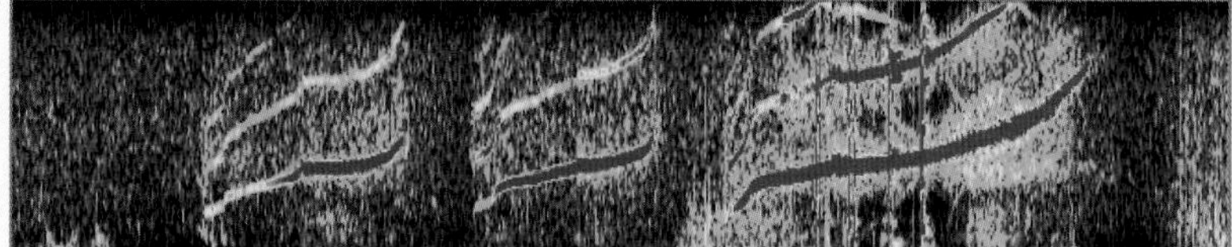

SYNTHETIC SIGNATURE WHISTLE

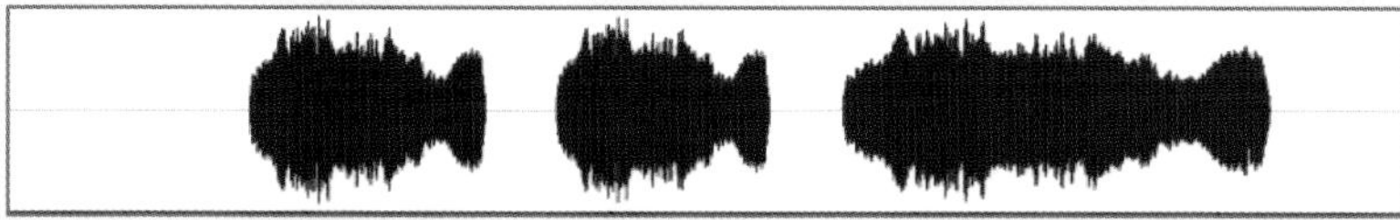

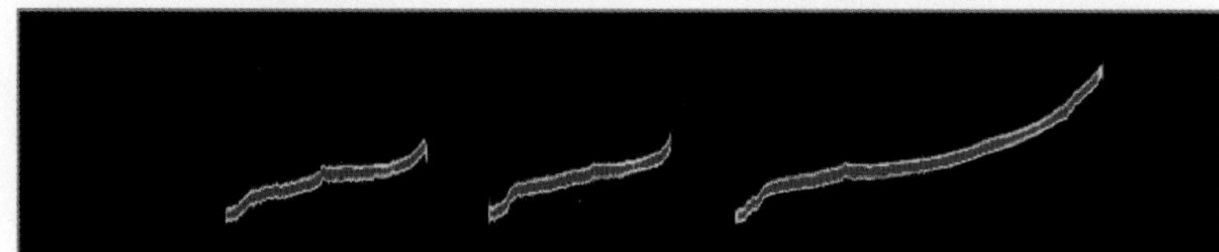

Left To test whether it is frequency modulation of the whistle that carries identity information, a computer was programmed to produce whistles that had all other voice features removed, including frequencies other than the lowest, or fundamental, frequency (multiples of the fundamental frequency are called "harmonics"). Shown here are spectrograms of sounds produced by a dolphin (top) and by a computer (bottom). Above each spectrogram is the waveform, which is a plot of loudness over time. Note that the different components of the natural signature whistle vary much more in their relative loudness than do those of the synthetic whistle.

3 SPECTROGRAMS OF SYNTHETIC WHISTLES
Synthetic whistle stimuli with all voice cues removed were played back to dolphins in the same way as were the natural whistles in the first set of experiments. Again, dolphins consistently responded more strongly to synthetic whistles of their relatives than to those of their close associates. A detailed analysis of signature whistle contours showed that related animals do not normally share modulation patterns, so these are unlikely to be used for recognizing relatives. Instead, these experiments confirmed that dolphins do use the whistle contour for individual recognition.

4 NON-SIGNATURE WHISTLES
In order to determine whether dolphins might also use voice cues for recognition, non-signature whistles were used as playback stimuli. These whistles do not have individually distinctive contours like signature whistles, but would still contain characteristic voice cues of the individual who produced them, if these cues exist. Unlike in the other two sets of playback experiments, dolphins did not respond more strongly to non-signature whistles of relatives vs. those of close associates, indicating that dolphins do not use voice cues to recognize one another's whistles.

5 INDIVIDUAL RECOGNITION
Although most, if not all, nonhuman terrestrial mammals use voice cues for individual recognition, dolphins appear to be unique in their use of the frequency modulation pattern alone to recognize the whistles of other dolphins. In fact, a detailed analysis of acoustic features of dolphins was unable to reveal voice cues in the modulation pattern of whistles. Perhaps changes in pressure during diving affects the sound production structures of dolphins, rendering voice cues ineffective. This then could have been a driving force behind the evolution of individually distinctive signature whistles, which are the closest analog to human names yet to be discovered in the animal kingdom.

CALL FUNCTION

Given the great diversity of cetacean species, with members ranging from the blue whale the size of three buses end to end, to the harbor porpoise, which only grows to the size of a human it can be difficult to find common themes in how they communicate. Let's look at what we know about call function a little more closely by tying together information about life history patterns, social structure, social behavior, and communication. There is more similarity than we might expect, with many sounds functioning to bring animals together.

MYSTICETES

The life histories of many mysticete whales are highly influenced by their long, annual migrations from polar feeding grounds to tropical breeding grounds. Some of these migrations are routinely more than 6,000 miles (10,000 km), with the longest known being more than 13,500 miles (22,000 km). It is not difficult to imagine that undertaking a trip of this magnitude on an annual basis would impact just about every aspect of an animal's life.

Indeed, migrations do constrain many important aspects of mysticete biology, such as the length of gestation, timing of and size at birth, and age at weaning. Typically a mysticete calf stops nursing at approximately six months of age, in time for the next migration to the feeding grounds. Mothers and calves do not appear to routinely associate together after the first year of life. In fact, there is little evidence for lasting, individually specific relationships among mysticete whales, and correspondingly, no evidence for mysticete vocalizations functioning as individual or group identifiers. However, relatively stable associations among pairs and trios of humpback whales have been documented in the Gulf of Maine off the east coast of the United States; an interesting area for future research would be to closely examine the vocalizations of these stable associates, to see if their communicative signals could be facilitating their relationships.

WHAT WE KNOW ABOUT WHALE COMMUNICATION

As mentioned above, our knowledge of mysticete whale communication is highly limited due to logistical difficulties in documenting senders and receivers of signals, as well as the effects of signals on receivers, when dealing with largely invisible animals and potentially vast scales of signal transmission. Although much research has focused on trying to elucidate the function of humpback whale song via behavioral observations and playback experiments on breeding grounds, remarkably, after several decades there is still only limited information regarding precisely how it functions.

However, there is strong evidence that humpback song, as well as the songs of other mysticete species, function as reproductive advertisement displays, akin to bird song. This assumption is supported by several observations. First, singers have been found to be male in cases where the sex of singing individuals could be determined, similar to birds that sing for reproductive advertisement. Second, and again similar to birds, song is usually produced on the breeding grounds, although it has been recorded in other contexts as well. Unlike bird song, which is generally well understood with respect to how it functions to maintain territories and attract mates in various species, it is not known whether whale songs are directed at other males, or at females, or at both.

As for other mysticete sound types, several have been described as likely "contact calls," such as the 20 Hz pulses of fin whales and right whale up-calls, and others have been associated with foraging behavior, such as blue whale "D" calls. Right whale "gunshot" sounds were found to be produced by mature males and were also hypothesized to play a role as a breeding display. However, most of these are generally anecdotal descriptions; little is known about how non-song calls function.

Song and mating calls are clearly key components of mysticete communication, which matches mysticete life history and social structure. Long-distance migrations likely limit opportunities for complex and stable social relationships and may require long-range

Above The sounds of mysticete whales are poorly understood, in part due to the large spatial scales across which they travel, making it very difficult to document who is producing the sounds and who its intended recipients are. Shown here are humpback males competing for a female.

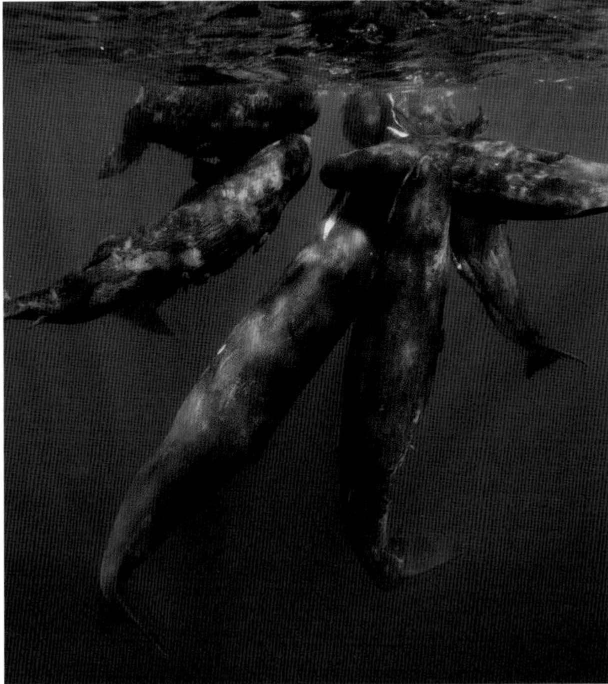

Above A pod of resident killer whales (left) in Johnstone Strait, Telegraph Cove, British Columbia, Canada and (above) sperm whale calves socializing off the coast of East Africa. Both killer whales and sperm whales produce sound repertoires that may facilitate group recognition.

communication systems to attract mating partners. Long-distance communication and its problems are discussed later in this chapter.

ODONTOCETES

In contrast to mysticetes, many odontocete species are characterized by having stable group or individually specific social relationships. Many species produce group or individually specific vocalizations, which are likely important in maintaining long-lasting bonds.

KILLER WHALES

Fish-eating killer whales offshore Vancouver, British Columbia, live in the most stable groups known for any mammalian species. Every animal stays in its matrilineal group, which is the group into which it was born. In most mammals, either males or females leave their birth group when they grow up, but in killer whales they stay. Their group-specific dialects may facilitate this stability. But it is unclear whether dialects really have a function; they may just be a result of animals copying each other. As killer whale pods grow larger, they eventually split and two separate pods develop. These pods then spend less and less time together, which leads to the slow development of vocal differences. Animals that split a long time ago may have very little in common in their call usage, while newly formed groups are very similar to one another. Thus, the degree of similarity in call repertoires could be useful to a whale choosing a mating partner, so it could avoid mating with a close relative (inbreeding). However, we do not know whether killer whales use this information in call dialects.

SPERM WHALES

Long-term, stable associations have also been described in sperm whales. The stereotyped patterns of clicks, called codas, produced by sperm whales have a communicative function, and were initially proposed to serve as individual signatures. Later studies found evidence for shared coda types, coda dialects, and large vocal clans, seeming to contradict the idea of an individually specific function. However, several studies now indicate that sperm whale codas may serve to identify both individuals and groups. Given that sperm whale social structure consists of stable matrilineal social units in which individuals have preferred associates, it is logical that their communication could facilitate both group and individual recognition.

Above A resident bottlenose dolphin in the Moray Firth off the coast of Scotland with a salmon.

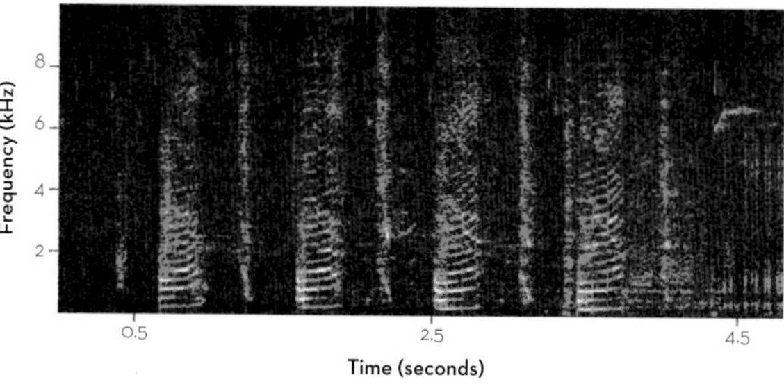

Above The bray call shown here is produced by bottlenose dolphins when they catch large fish. These calls sound like a donkey's bray and are much lower in frequency than other dolphin sounds. Other dolphins rush toward a braying individual, but it is unclear whether the sounds evolved to scare the fish or to attract other dolphins.

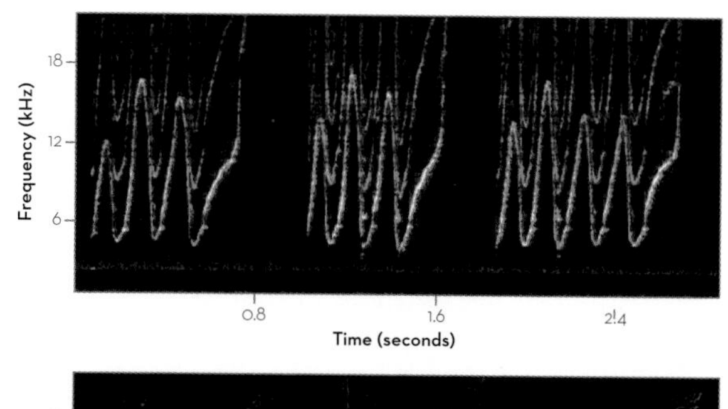

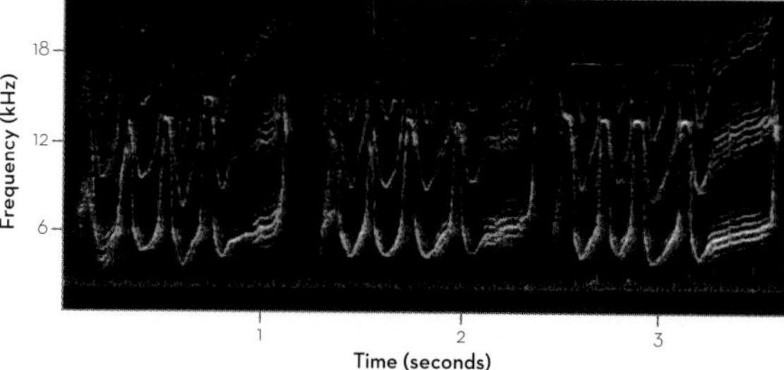

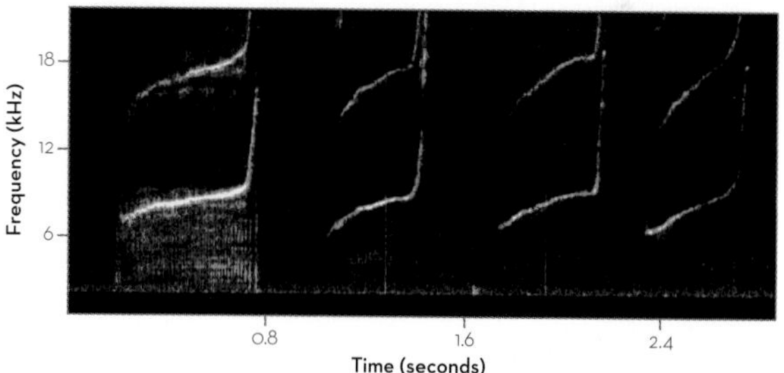

Above Bottlenose dolphins sometimes copy one another's signature whistles, which may be a way to call another dolphin by "name." Here the upper and lower panels show spectrograms of signature whistles of two associated dolphins (A and B), and the middle panel shows dolphin B copying the whistle of dolphin A.

BOTTLENOSE DOLPHINS

Bottlenose dolphins live in "fission-fusion" societies (see Chapter 5), which are characterized by strong individually specific social relationships superimposed on a larger network of more fluid relationships.

As we saw above, bottlenose dolphins develop their own signature whistle early in life, although the process by which a young dolphin "chooses" its signature is not understood. They seem to learn parts from other dolphins, but also modify what they learned and thus invent their own unique contour. We also know that dolphins maintain vocal flexibility throughout their lives. This allows them to copy signature whistles of others, the function of which seems to be to call other dolphins when they need to get in touch with them, similarly to how we use names. Signature whistles are key to dolphins' long memory for social bonds; they are able to recognize signature whistles of past associates over periods of more than 20 years.

The large variety of non-signature whistles and burst-pulsed calls produced by dolphins are in great need of study. Almost nothing is known about how these calls function in the dolphin communication system, with a few exceptions. "Pop" vocalizations were found to be associated with male alliances in consortships with females, and were hypothesized to be threat vocalizations used to induce females to remain close (see Chapter 5). Female consorts turned toward males at higher rates when males produced pops, and aggressive "head jerks" were also associated with pops. Other vocalizations, "brays," were recorded while dolphins were feeding on salmonid fish, and were found to co-occur with overlapping matching whistles, suggesting a social aspect to food-calling behavior in dolphins.

CASE STUDY: **SIGNATURE WHISTLE FUNCTION & PERCEPTION**

The previous case study describing playback experiments (see pages 78–9) showed that dolphins use signature whistles for individual recognition. But how do dolphins actually use these whistles in their daily lives? A study of a group of captive dolphins sheds some light on this question.

1 INITIAL SET-UP

Four captive dolphins that lived in a dolphinarium at Duisburg Zoo, Germany, were selected for study. The dolphins—Delphi, Duphi, Pepina, and Playboy—were housed in a tank with two separate pools. Two underwater microphones, known as hydrophones, were set up, with one in the main pool, and one in the side pool.

2 VOLUNTARY ISOLATION

Researchers observed behavior of the four dolphins and noted when an individual voluntarily went into a small side pool that contained one of the hydrophones. From their recordings, they identified which whistles occurred in the side pool, and compared these whistles to those that occurred in the main pool.

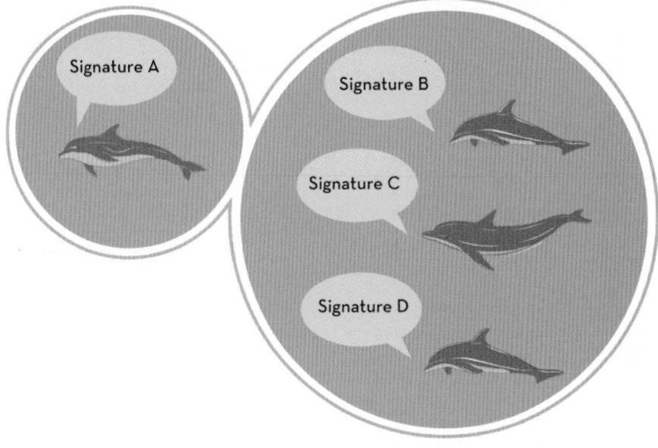

3 INDIVIDUAL RESPONSES

Signature whistles almost always occurred only when an individual was by itself in the side pool. In almost all cases, the signature whistle corresponding to the

DELPHI (SIGNATURE WHISTLE A)

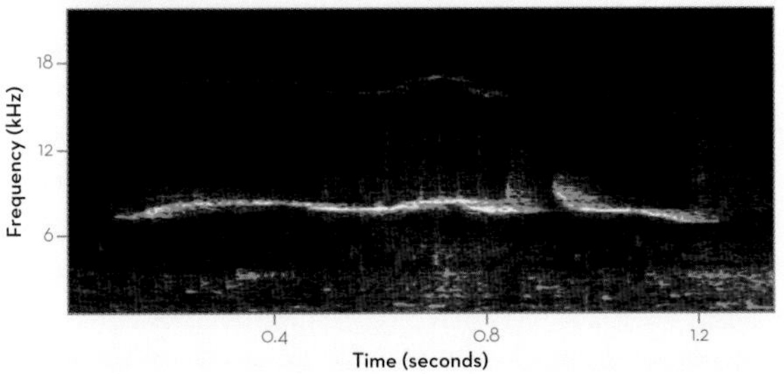

PEPINA (SIGNATURE WHISTLE B)

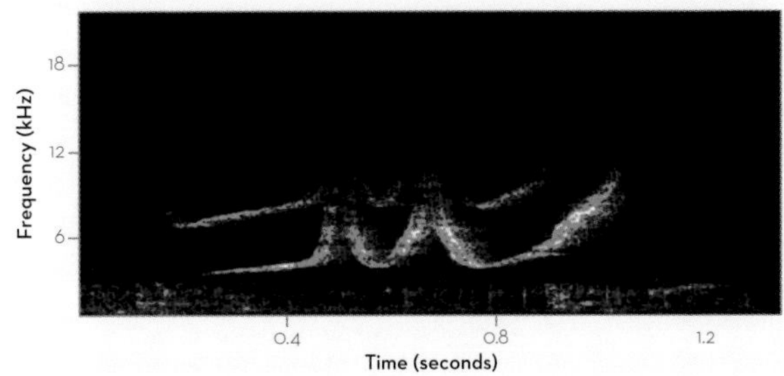

DUPHI (SIGNATURE WHISTLE C)

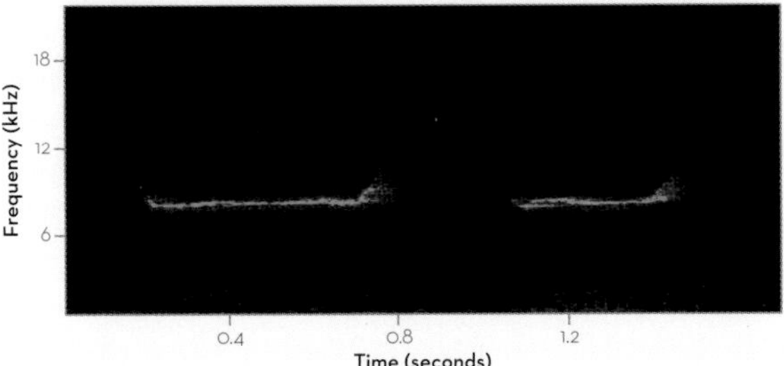

PLAYBOY (SIGNATURE WHISTLE D)

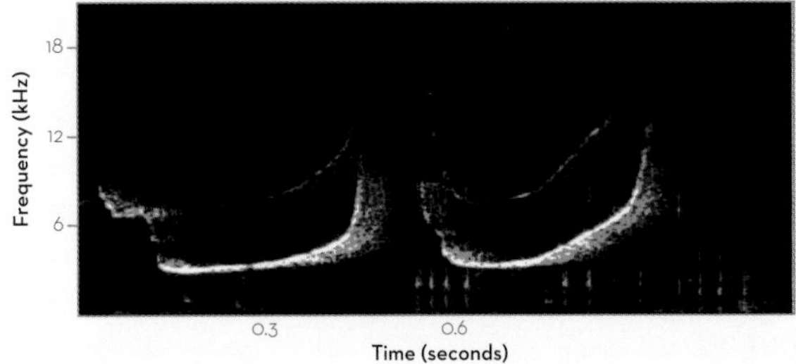

Above Spectrograms of the signature whistles of the dolphins in 1995. Each animal produced primarily its signature when it was out of sight of the rest of the group.

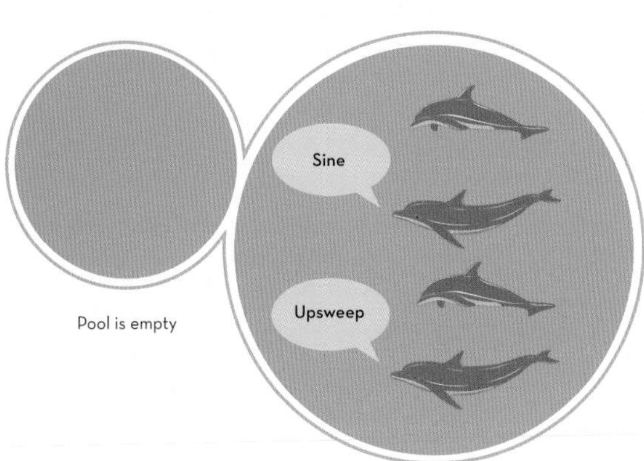

Left Dolphins had access to all pools at all times. When one animal was out of sight (far left), all four used their signature whistles. But when swimming in plain sight of each other in one pool (left), they did not use their signature whistles but other communication signals.

isolated animal was recorded from the side pool while the other animals in the main pool primarily produced their own signature whistles in response, suggesting that animals were trying to maintain contact with each other while one of them was away from the group.

4 NON-SIGNATURE WHISTLES

In contrast, few signature whistles were recorded when all animals were together in the main pool. Instead, a variety of non-signature whistles (such as generic sine and upsweep-shaped whistles) were typically recorded in this context.

5 OUTCOME

This study demonstrated that signature whistles are used for maintaining group cohesion when animals are separated from each other.

VOCAL LEARNING

Throughout the previous sections, we highlighted vocal learning capabilities of cetaceans. The ability to learn how to produce sounds based on auditory input is called "vocal production learning." As humans, we take this ability for granted, as it is so central to the development of human speech. However, this ability is surprisingly quite rare among nonhuman mammals. Even our closest relatives, the nonhuman primates, are born more-or-less "hard wired" to produce their species-specific vocal repertoire. In contrast, many marine mammals are capable of vocal production learning. Why?

SOCIAL RECOGNITION

It seems that vocal learning is used in different contexts among cetaceans. One of these is social recognition. Most, if not all, terrestrial mammals are capable of identifying one another by means of voice characteristics, also called voice cues. Since these cues are influenced by the shape and size of the sound production organs, they would likely be affected in water by pressure, which changes with depth. The lack of reliable voice cues could have been a driving force behind the evolution of vocal learning in some cetacean species. This explanation makes sense for bottlenose dolphins, which clearly have strong individually specific social relationships that require a robust mechanism for individual recognition. With visual cues being of limited usefulness in turbid marine environments, and with the reduced ability of dolphins to detect olfactory cues, an acoustic cue is the obvious choice (see the case study on playback experiments on pages 78–9).

When bottlenose dolphins copy the signature whistles of close associates, it appears to serve an affiliative function, helping to maintain contact between mothers and calves or between adult alliance partners. Copies are imperfect, such that they are likely recognizable as copies, thus not affecting the capacity of signature whistles to serve as individual identifiers. Therefore, recognition for a highly social animal in a very homogeneous environment such as the sea is a likely candidate for the origins of vocal learning. But what about vocal learning in less social species?

ATTRACTING A MATE

All singing mysticetes show a pattern very similar to that of birds, which learn their songs, where males sing to attract females and to repel other males. So this could have been another reason for the evolution of vocal learning in cetaceans. Yet the situation is not that clear for all song aspects. The changes in humpback whale song that occur from year to year, which all males match, are clearly derived from vocal learning. However, the function of these changes is not understood. It could be that males have to synchronize their song to get females interested in mating, as can be found in some terrestrial animals, but this is mere speculation at this point. The pattern of learning from west to east in the southern hemisphere is even more puzzling (see Chapter 6).

RECONSTRUCTING EVOLUTION

Vocal learning is clearly common in cetaceans. There is strong evidence for a role for learning in signature whistle development and copying in bottlenose dolphins and in the production of group-specific calls by killer whales. Killer whales, Risso's dolphins, and bottlenose dolphins have been found to learn call parameters from other species if they are housed with them. Beluga whales and bottlenose dolphins can match the rhythm and some tonal qualities of human language when in human care. In more controlled experiments, bottlenose dolphins have been successfully trained to copy computer-generated whistle sounds. Recognition or song learning could have been the first context in which vocal learning was used, or it may have evolved in a different context from which it is used today. The reconstruction of behavioral evolution is difficult since behavior patterns do not leave fossil records.

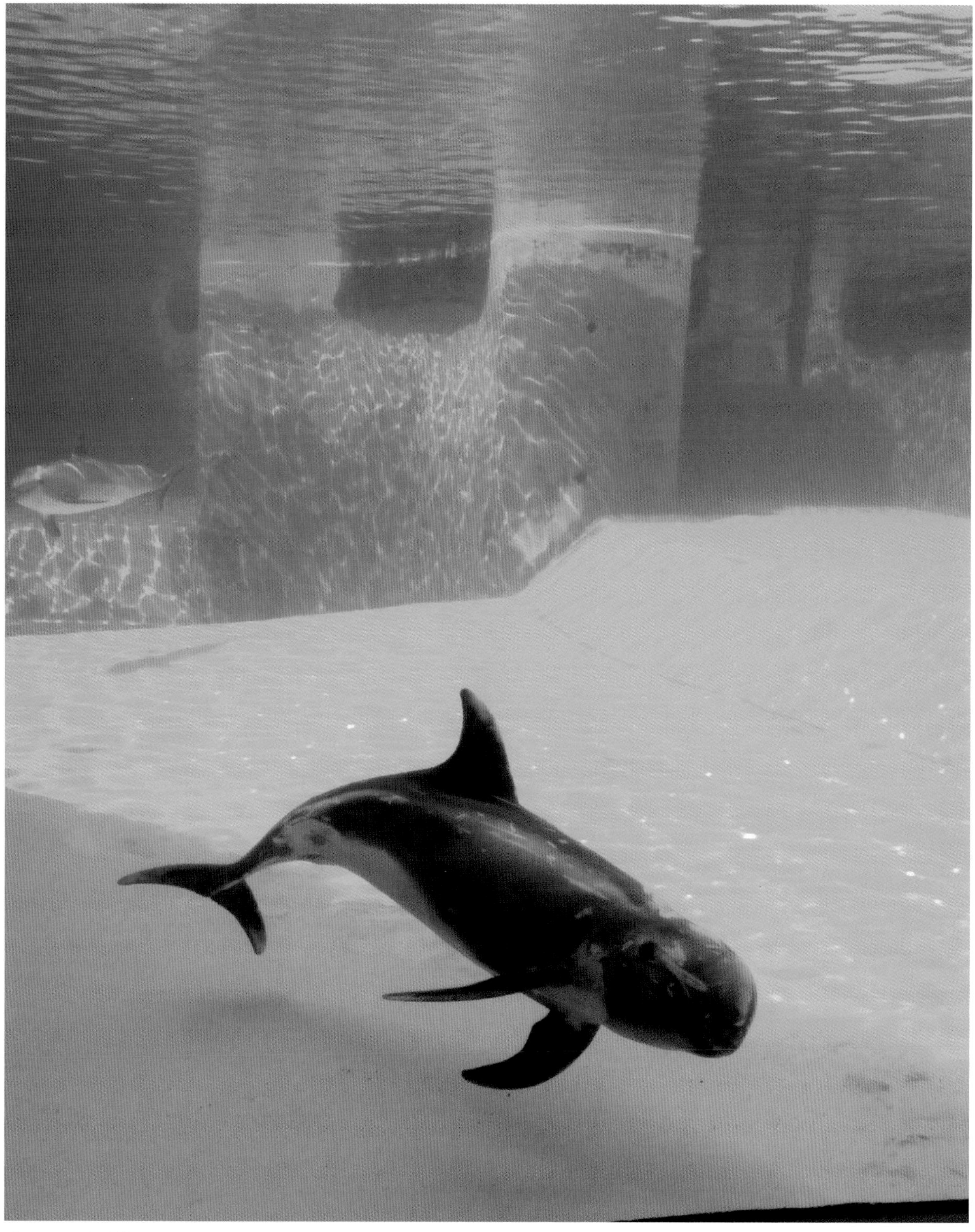

Above It can be difficult to determine which vocal signals are learned vs. "hard wired" in an animal's repertoire. But when an animal produces sounds similar to those of another species, the signals must be learned. An orphaned Risso's dolphin in captivity learned call parameters from bottlenose dolphins with which it was housed, providing evidence of vocal learning.

Below Common dolphins off the South African coast. These dolphins are not often seen in aquaria but were the second species in which signature whistles were found by researchers in the 1960s.

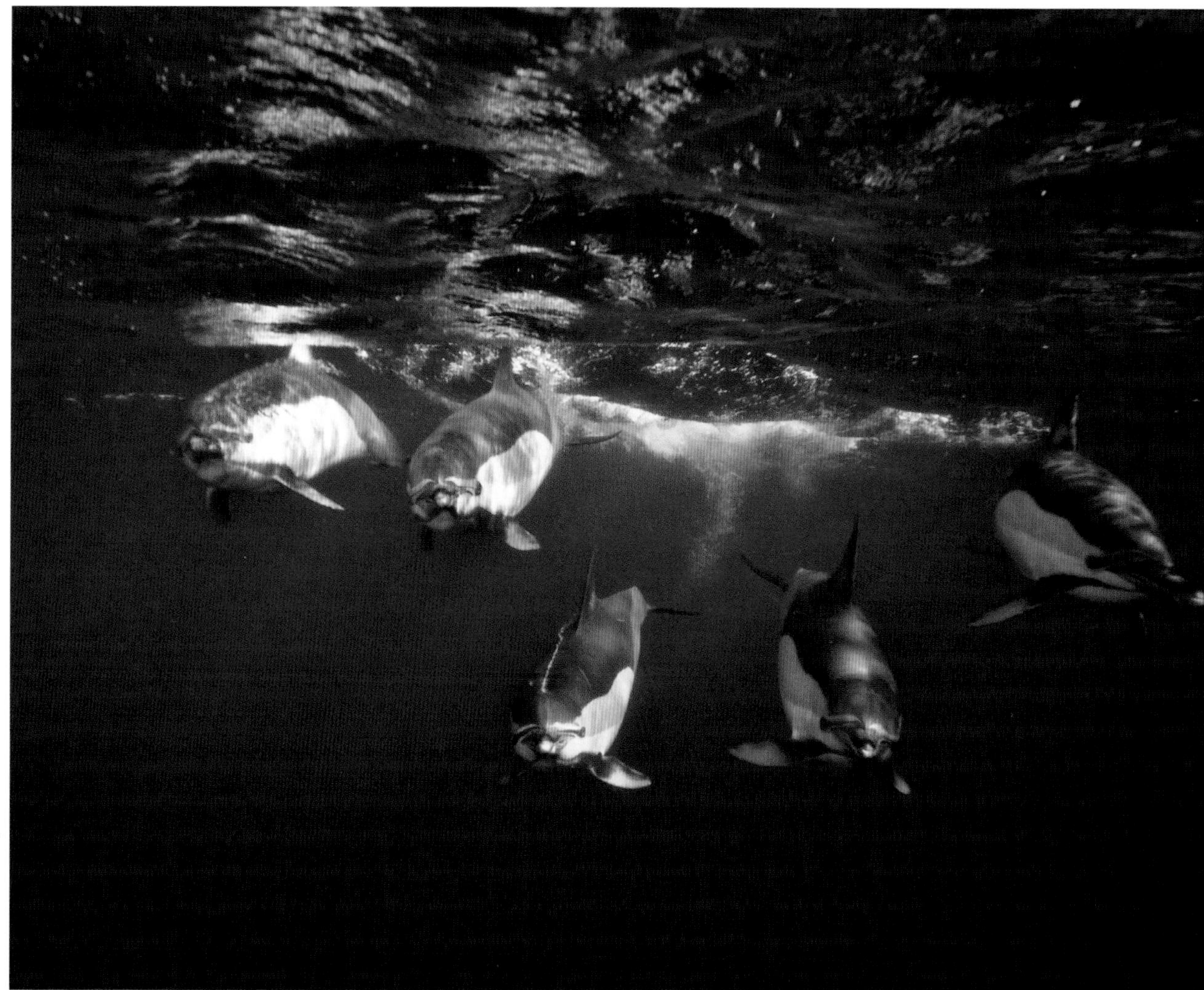

WILL WE EVER BE ABLE TO TALK TO DOLPHINS?

Dolphin intelligence is a lively field of research that has uncovered impressive abilities, sometimes rivaling those of the great apes. Bottlenose dolphins are able to imitate the behavior and vocalizations of others, they can form concepts, understand syntax, have long-term social memory, cooperate when hunting, and understand as well as use referential communication (that is, signals that refer to objects or individuals). The cognitive abilities of dolphins have led some to believe that dolphins have a communication system as complex as ours and that we might be able to establish some sort of contact with them. Several scientists, including John Lilly, mentioned at the beginning of the chapter (see also Chapter 3, which discusses his work), tried to teach dolphins to communicate with humans, but none of these efforts led to much success. Dolphins tend to use the signal for "fish" a lot, but do not seem to engage in more complex interactions using such signals. More success was found in experiments with great apes, but they are much more similar to us than are dolphins. Dolphins' experience of the world is primarily acoustic, not only through communication but also through echolocation. We have not had any shared evolutionary history with whales or dolphins in the last 60–70 million years. Since then, their brains

Below Spinner dolphins live in large groups and use whistles to communicate when they are far apart and burst-pulsed sounds when they are closer together.

have experienced changes that are unlike the ones in our own evolutionary history. The ways that dolphins process information and experience the world are likely to be very different from ours. The reasons that humans think of them as kindred spirits are primarily based on their curiosity and willingness to interact with the world around them. But they show such curiosity toward fish and other animals they encounter, as well as toward humans.

To understand dolphin minds, we must remain open to the discovery of new forms of communication, not simply the ones we use in human language. If we focus too much on language, we are likely to miss other aspects of their communication that may be more important to dolphins than reference, syntax, or word use. Imposing our own communication system by teaching them to speak is unlikely to be a successful avenue. For example, there is little evidence for syntax playing an important role in dolphin communication, yet it is paramount in ours. In training, we may be able to teach dolphins a syntactical system, but this could push them into using signals in a way that is unnatural to them. Studying how dolphins use communication signals in their daily lives promises to be the best avenue toward a deeper understanding of their natural communication system and how they see the world.

LONG-DISTANCE COMMUNICATION IN A NOISY WORLD

Communication does not happen in isolation between two animals but usually involves anyone within earshot. The range over which cetaceans can hear each other is much larger than for any terrestrial animal, making their communication networks very complex. But the improved sound transmission at sea also makes other sounds a more serious problem for them. Their communication signals can be masked both by sounds produced by other animals and noise produced by human activities.

Ever since zoologists Roger Payne and Doug Webb published their paper "Orientation by means of long range acoustic signaling in baleen whales" in 1971, people have been fascinated by the idea that whales may communicate across ocean basins. Payne and Webb showed that it is theoretically possible for the very low frequency sounds produced by both blue and fin whales to travel distances of thousands of miles (low frequency sounds can travel greater distances than higher frequency sounds). Both blue and fin whales often produce repetitive calls that are characterized by frequency sweeps, which are features that could facilitate long-range transmission. Although broadcasting sounds across ocean basins may be theoretically possible, more realistic estimates of the functional range of fin whale calls are on the order of 55 miles (90 km), which make more sense from a biological perspective: animals hearing a sound over many hundreds of miles are unlikely to reach a calling whale in a biologically relevant time frame.

However, documenting changes in behavior over scales of even less than a mile is still daunting. For the most part, it has only been possible to make very general associations between call types and behavioral states for mysticete whales. So far, the vast majority of mysticete call types are known only by descriptions of call parameters and correlational behavioral data. Much work remains to be done to understand how these sounds function in their natural communication systems.

COMMUNICATION NETWORKS

The classical way of investigating communication is to look at the interaction between two animals, a sender and a receiver. However, this is not always the most realistic scenario. Each sound can be heard over a certain area, and all animals within that area will be able to eavesdrop on a signal. This means that there are a number of animals that form a communication network, rather than just two interacting individuals. These networks are particularly large in aquatic environments such as lakes or oceans. The reason is that sound loses much less energy in water than in air, with the result that sounds can be heard over larger distances in water. At first glance this seems to be a big advantage for cetacean social networks. Being able to communicate over vast distances would seem to make it easy for cetaceans to stay in touch. However, there is a flip side to that coin. Imagine you could hear everyone talking within 10 miles (16 km) of you right now. Noise created by other organisms is a serious issue for communication in the marine environment, and it becomes more complicated as the density of animals in an area increases. As mentioned above, fin whale sounds, for example, can be heard over tens of miles, but on average there is only one fin whale or less in each 40 square miles (100 km^2). Dolphin sounds, on the other hand, are maybe detectable over "only" 6 miles (10 km), but there may be hundreds of animals in the same area. Thus, dolphin communication networks are often much larger than those of mysticetes, even though dolphin sounds do not travel nearly as far.

The impact of noise on cetacean communication

Noise abounds in the oceans, from natural sources and man-made ones. Some of the anthropogenic contributors shown here are seismic vessels with airguns looking for oil reservoirs below the sea bed, and engine noise from vessels such as the cargo ship and submarine. Other sources not shown include oil drilling platforms and military ships that produce very high intensity sonars. Sound intensity is given in decibels (or dB). All values given here are maximum levels in reference to 1 microPascal, the typical unit for these measurements underwater. They are not comparable to decibels measured in air.

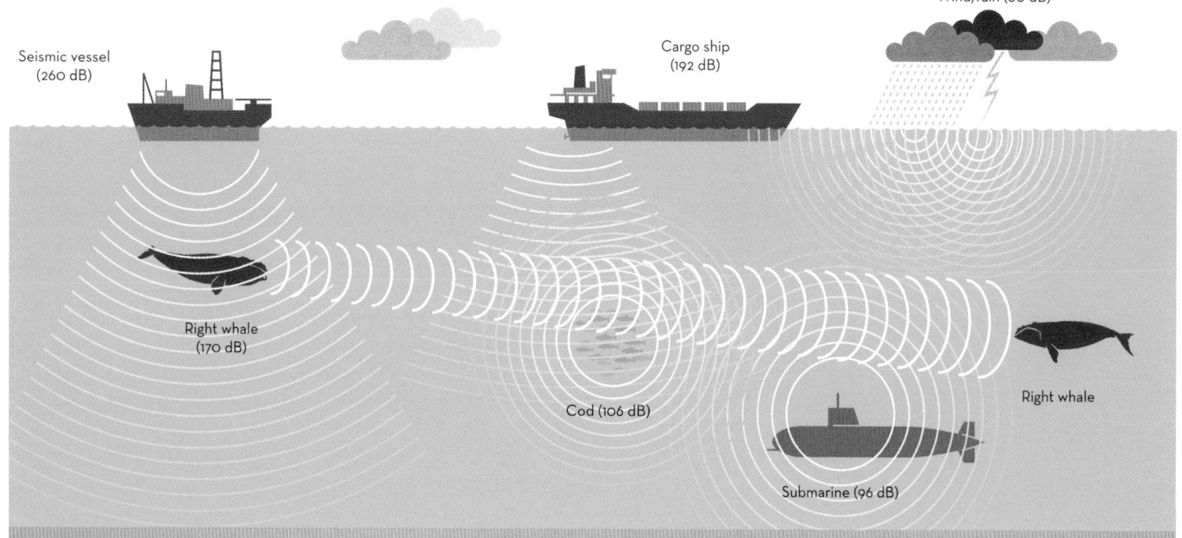

It is still unclear the extent to which cetaceans may use information from eavesdropping. We know that dolphins that are close together can eavesdrop on the echoes of each other's echolocation clicks. But it is less clear whether more distant information has much impact on dolphin behavior. Mothers and calves of many cetacean species become separated at times, but it is rare to find them more than about two-thirds of a mile (1 km) apart. Over this distance, animals can communicate effectively and are able to stay in touch. Sounds from more distant animals may indicate good foraging spots or mating grounds, but more detailed information is unlikely to be transmitted over long ranges.

THE CURSE OF HUMAN NOISE

No matter whether we are talking about ranges of 100 yards or 100 miles (or indeed 100 meters or 100 kilometers if you prefer measurements in metric), cetacean communication has great potential to be impacted by human noise. Just like communication signals, noise from human activities travels far in the ocean and contributes to the background noise over which cetaceans are trying to stay in touch. Thus, our activities can lead to the disruption of communication networks and consequently may negatively impact many aspects of the lives of whales and dolphins.

In its most extreme form, noise may frighten animals and lead to panic that could result in death, or it may cause stress that could result in more chronic effects. Subtle effects of noise include masking of communication sounds, or deterioration of hearing, both of which could impact animals' abilities to maintain contact with others at sea.

Given the importance of acoustic signals in their lives, noise has great potential to impact cetacean survival and reproductive success. The main sources of human noise are shipping, marine construction (for example, wind farms or oil platforms), marine exploration (using airguns to map subsea structures or to find oil), and sonar (particularly those used for detecting submarines). In heavily used areas such as outside large harbors, noise can become overwhelmingly loud, but even in the open ocean, noise from distant sources can accumulate and disrupt cetaceans. Applied research on effects of increasing levels of noise on cetaceans is urgently needed, and for this we need to expand our basic knowledge of how and why they communicate.

5 QUINTESSENTIALLY SOCIAL CETACEANS

Janet Mann

Humpback Whale Portrait 1341
Kingdom of Tonga, 2006

"While photographing this young whale and his mother, I momentarily lost track of the calf's position. In that moment, I felt a body rest against my back. I looked up and saw the young whale's chin settle over my head. The whale then wrapped its pectoral fin around my body and held me close for a few moments"—Bryant Austin

A CLOSE FRIENDSHIP

Cetaceans are among the most sociable creatures known in the animal kingdom. Many species, especially odontocetes, not only group together but also protect each other and form long-term social relationships. Furthermore, the complexity of their social behavior and bonds rival or even exceed those of most primates, which raises critical questions as to why such intricate social communities have evolved in cetaceans? Members of some cetacean species interact with hundreds of individuals, much larger groupings than any primate. At the same time, mother–offspring bonds are intimate and endure over years, even decades. In some species, stable cooperative relationships between unrelated individuals are common, a facet of social life rarely found outside of humans. Here we explore the varied and fascinating social lives of dolphins, porpoises, and whales.

It was a traumatic day, March 19, 1994 in Shark Bay, Western Australia. Hobbit, a four-month old female bottlenose dolphin calf, was meandering about 75 yards (70 m) away from her mother Holeyfin, when a 6 ft (1.8 m) tiger shark attacked, biting off Hobbit's tail flukes, then disemboweling her quickly. Holeyfin's adult daughter, Nicky, and close female companion, Surprise, sped to the scene and fought the shark alongside Holeyfin, risking their own safety. Moments later, the fight was over and Nicky swiftly escorted her young son away from the area, while Surprise and Holeyfin remained with Hobbit. Surprise rested her pectoral fin on Holeyfin's side as the frantic mother pushed Hobbit's carcass to the surface. Holeyfin, who almost never whistled, now whistled constantly—easily audible in air. Surprise stayed close to Holeyfin for another half hour before swimming off—but, at a few hundred yards, she was still close enough to respond to Holeyfin if needed. Indeed, the tiger shark had not gone far but was now in the seagrass bed—less than a hundred yards away, while Hobbit had drifted into shallow sandflats near shore. My student and I were with Holeyfin—and had quickly developed a sampling protocol to record these events systematically, but we could not capture the emotional tenor, Holeyfin's frantic behavior and attempts to resuscitate her calf. Suddenly Holeyfin darted off from Hobbit and was thrashing in the water. I first thought she was again attacking the shark, but she quickly emerged with an enormous fish in her mouth and swallowed it whole. I admit to being really confused. How could she hunt and eat at a time like this? I did not quite understand her emotional state or her relationship to Hobbit (whom she was rather neglectful of since birth). In fact, the shark returned to finish off Hobbit, but Holeyfin, who was now looking for her next meal, sped back in time (this time) and chased the shark off. The sun was now set and with darkness falling, I decided to pull Hobbit's carcass from the water so we could find out why, prior to the attack, she had sores on her tiny body and was emaciated. I also worried about Holeyfin spending the night defending Hobbit as more sharks were sure to come. As I pulled Hobbit into our dinghy, Holeyfin pushed Hobbit up one last time, coincidentally assisting me. Once Hobbit was in the boat, Holeyfin left the area immediately. A postmortem revealed that Hobbit had broncho-pneumonia, but we never found out what the sores were from. The next two days, Holeyfin returned to the site of the attack and where Hobbit drifted— whistling nonstop. She spent most of her time with Surprise and her young grandson Rabble. By the third day, her behavior appeared to be normal and males were already becoming interested in fathering her next calf.

Opposite top Holeyfin guarding her dead daughter Hobbit, who is floating belly-up at the surface. This was Holeyfin's last calf, as she died the following year when a stingray spine pierced her heart. Holeyfin was survived by two daughters, three grandoffspring, and two great-grandoffspring.

Opposite bottom Shark Bay has one of the largest tiger shark populations known and about 75 percent of the dolphin population bears shark bite scars.

INTERPRETATIONS OF SOCIAL BONDS

While scientists rarely see such events, they tell us much about the importance of social bonds, but highlight perplexing behaviors as well. This event underscored Holeyfin's close relationship with Surprise, which had been evident for years by their frequent association and close physical contact. Remarkably, Holeyfin would often defend Surprise when she was attacked by Nicky, Holeyfin's own daughter. (Aggression is rare among female dolphins, but Nicky was exceptional in this regard.) And it shows a mother's dedication to defending her offspring, even if she failed to prevent the original attack. Holeyfin's distress, evident by her constant whistling, attention to Hobbit's body, and returning to the site of the attack, is something we understand. Defending one's kin and allies is what we would expect. But there were some bewildering aspects as well. The fact that Holeyfin interrupted her vigilance to catch a fish, when the shark was nearby, is alien to us, and probably other primates, since we usually forgo food in such circumstances. She also "seemed" normal within a few days, except that her four-year-old grandson seemed particularly interested in her, but not in the way we might imagine, since he mounted Holeyfin repeatedly, dozens of times. He wouldn't necessarily "know" Holeyfin was his grandmother, but would know she had just lost her calf. She would know who he is. The grandson was far too young to be fertile, and mother-son matings are common when the calf is young (still nursing), but this was the first grandmother-grandson pairing we had seen. Such behavior is yet another facet of social life that eludes us, a species with strong incest taboos.

UNDERSTANDING CETACEAN BEHAVIOR

The early scientists of animal behavior (ethologists) developed a term for the experience of another organism—umwelt. Perhaps nothing is more elusive than understanding how another organism might experience their social world, from encounters with strangers to regular interactions, relationships, and long-term bonds with kin, non-kin, and mates. And, of course, the development, maintenance, and loss of close bonds are part of that fabric. As an obligatory social species that is capable of great empathy, we believe we understand the emotions that accompany social living—at least for fellow humans. In recent years it has become less taboo for scientists to speculate on the emotions of other animals, in large part because we now recognize the depth and importance of social bonds for so many species, in addition to figuring out better ways to tap their umwelt experimentally.

With or without speculation on emotional state, from the scientific perspective, emotions are hypothesized to support survival and reproduction. Fear, pain, anxiety, loss, depression, joy, pleasure, and affection are states that correspond to important social interactions and relationships. In fact, whenever scientists examine the function of close social bonds, they find that affiliative relationships promote a myriad of benefits ranging from enhanced immune system, reduced disease risk, reduced predation risk, better defense against male harassment or aggressive conspecifics, social skill development, social support, information sharing, and enhanced survival and reproduction. So, while we might not understand the umwelt for animals with an environment and cognitive, sensory-perceptual, and motor abilities so different from our own, we can still define social contact, bonds, and structure and examine the benefits of such phenomena.

SOCIAL CONNECTIONS FOR SURVIVAL & REPRODUCTION

Our work on Indian Ocean bottlenose dolphins in Shark Bay has consistently shown that social bonds are good predictors of survival and successful breeding from the earliest life stages. For example, if a male calf's mother is well connected, he is more likely to survive the risky juvenile period after weaning and reach adulthood. Juvenile females with strong social ties have higher success in raising calves later in life than females with few or weak ties. In adulthood, those females who associate with reproductively successful females are themselves likely to be successful, meaning that their calves survive to weaning. This might be because such associations offered protection, information, and social experience for offspring. And, adult male bottlenose dolphins are unlikely to father any offspring unless they are members of a long-term stable alliance (pairs and triplets of males who repeatedly cooperate to gain access to females and in battle against other male alliances). Clearly, evolution has favored social living in dolphins. Social connections really do matter.

Opposite top Mother and calf killer whales engaging in a synchronous inverted breach (leap). Such synchrony is common between closely bonded individuals such as a mother and calf.

Opposite bottom Amazon river dolphins engaged in social behavior. This interaction appears to be friendly, but the open jaw could also suggest aggression. One needs to know the full context to interpret such interactions.

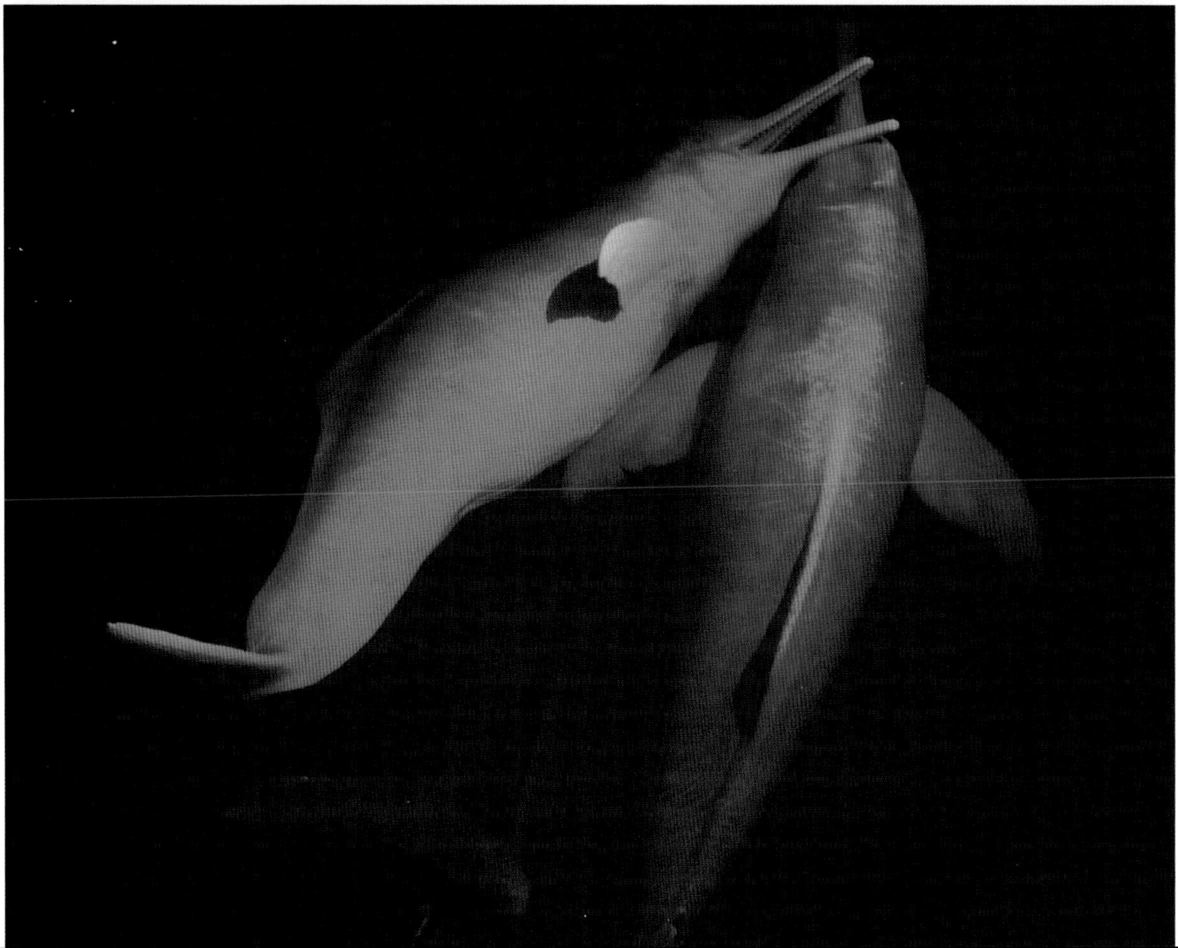

WHAT IS SOCIAL LIVING?

Many species may be described as living in social groups. These vary from anonymous aggregations (insect swarms, migrating herds of wildebeest), to highly differentiated social relationships such as those found in elephants, primates, spotted hyenas, and dolphins. While some animals group to use the same resource, such as food or refuge, or to reduce the likelihood of being eaten, others group primarily with specific individuals, whether they be kin, acquaintances, or allies. Such differentiated relationships are the hallmark of social complexity and are a key feature of many cetacean societies.

Social system, network, organization, and mating system all represent the emergent properties of encounters, interactions, and relationships among individuals. The durability and nature of social encounters helps to define whether a species is considered to be "social or group living" or "solitary."

What distinguishes social from solitary species is the degree to which they associate and coordinate their activities beyond mating and parental care. Obviously all sexually reproducing animals must encounter each other to mate, and all mammalian females nurse and have physical contact with their offspring. Since paternal care is absent in cetaceans, there are no pair-living family units. Some of the key factors driving the social organization and patterns of grouping are calf protection, predation risk, prey distribution, hunting tactics, and social competition. Information sharing and cultural transmission of foraging, ranging behavior, anti-predation tactics, and social practices are all important benefits of group living.

KIN GROUPS, CLANS & PODS

All odontocetes are social, meaning that individuals have frequent encounters with fellow members of their species beyond mating and maternal care. Some odontocetes also have extended kin groups that consistently spend time together. Sperm whales form stable matrilineal units that associate with larger clans, a pattern thought to be driven by calf protection. Sons emigrate when they approach sexual maturity and at full size might rove singly between female groups. This pattern likely benefits males, because staying with one matrilineal unit would limit his reproductive opportunities. And, females need not compete with large males for squid.

Killer whales form stable matrilineal pods that are "bisexually philopatric"—meaning sons and daughters remain within their natal area or with their mother for life, although the size of these units depends on hunting strategy. Fish-eating killer whales of the Pacific Northeast, for example, form large matrilineal pods where both sons and daughters remain with their mothers for life, with pod sizes as large as 100. They spread out to hunt individually, although food-sharing is common. Mammal-eating killer whales that are sympatric—or overlap geographically—with fish-eating whales form smaller units that enhance their hunting efficiency of harbor seals. Mammal-eating killer whale pods have three or four members, a mother, her sons, and dependent offspring. Her daughter will, by adulthood, form her own pod. Among mammal-eating killer whales, cooperative hunting by these small units is essential for capturing seal prey, which are always shared.

Strong, stable family units are also found among pilot whales, false killer whales, pygmy killer whales, and some beaked whale species—although the precise mechanisms driving their social structure are not known. Notably, many of these species specialize on squid; a number of cetacean biologists have proposed that deep-diving for squid may favor stable group-living units because individuals can share foraging information and reduce predation risk, especially for calves that are otherwise left alone at the surface during maternal foraging.

SOLITARY CETACEANS

At the other end of the spectrum are fleeting associations between many of the mysticete species. Because of this, most mysticetes, including blue, bowhead, Bryde's, fin, gray, humpback, pygmy right, right, and sei whales are classified as solitary. They might

Killer whale pod structure

Killer whale social network structure in the Pacific Northeast where the resident, fish-eating whales have been studied for over 40 years. The circles represent whales and the lines connect them to their preferred associates. Each cluster represents a matrilineal unit or pod (natal units). Some of the pods are closely connected. After Williams and Lusseau (2006) by permission of the Royal Society.

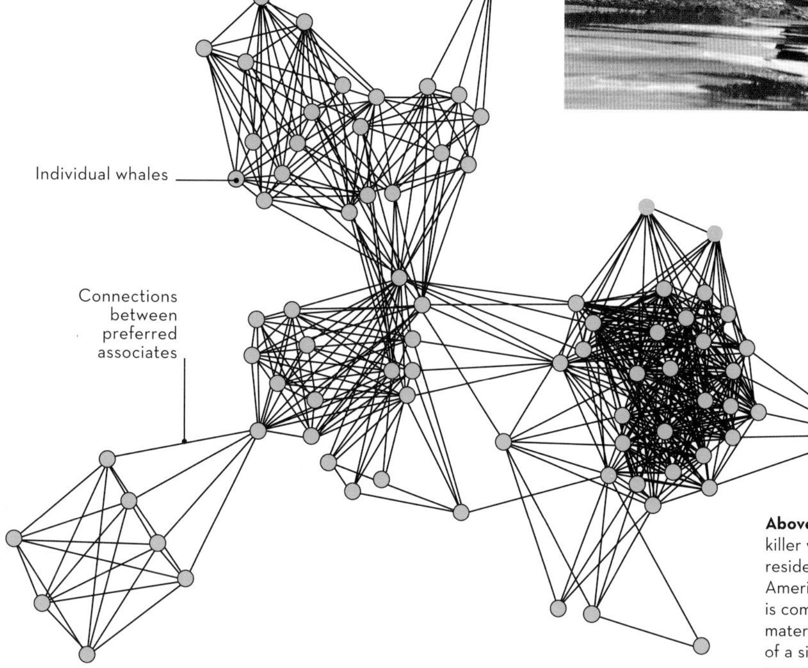

Above Much of what scientists know about killer whale societies comes from studying the resident pods off the Pacific coast of North America. The basic social unit, or matriline, is composed of individuals connected by maternal descent and may include members of a single generation or up to as many as four generations. Matrilines are highly stable and individuals may not stray from the family group for more than a few hours.

encounter the same associates at preferred feeding and breeding sites, and have social relationships, but most relationships are ephemeral in nature. Mysticetes that have more transitory contacts where individual recognition is less important might still be found in groups, but they are not in what is considered socially bonded groups, with well-differentiated long-term bonds. There are some important exceptions to this in the better-studied mysticetes, such as humpback whales, begging the question as to whether it is our limited perspective, rather than the whales, that defines their social world.

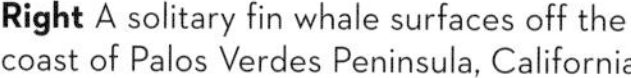

Right A solitary fin whale surfaces off the coast of Palos Verdes Peninsula, California.

SOCIAL DIVERSITY

Cetaceans exhibit enormously diverse types of social organization, even within species, and even within one location. For example, bottlenose dolphins might be resident to an area (individuals maintain the same geographic location year-round), while other members of the same population are migratory, interacting with residents on a seasonal basis. Groups of common dolphins can number in the thousands when far offshore (known as the pelagic zone, or open sea), presumably because they are feeding on large prey schools, sharing information about prey, or reducing predation risk, but might reliably be found in much smaller groups (less than ten individuals) in more coastal locations. Mysticetes are classified as solitary but they sometimes aggregate at breeding grounds or feeding grounds in very large numbers. Temporary associations also form during the migration. Males are known to accompany female humpback whales and their offspring during the migration. Males have even been observed protecting calves from killer whale attacks. Such associations over weeks or perhaps months might be mutually beneficial in that the male has a better chance of mating with the female and the female garners extra protection for her calf. These associations are interesting because adult whales are not particularly vulnerable to predation, but might opportunistically group to protect their offspring.

Among the most pelagic dolphin species that tend to form large aggregations, such as common dolphins, both sexes might move between groups or communities, so relatedness with other members of the aggregation tends to be low. It is difficult to characterize these large groups as they might have much more structure than what first meets the eye. For example, even in large aggregations of hundreds, there might be smaller subgroups that are staying together. Given the challenges of differentiating individuals that are not terribly distinctive, are moving quickly, and number in the hundreds, it is no surprise that scientists have not really studied their social structure (who associates with whom) except by looking at genetic structure. Since the samples come from strandings or biopsies of individuals in the larger population, not necessarily individuals that associate closely, an underlying structure might be obscured. Interestingly, studies of common dolphins, long thought to have low kin-relatedness due to dispersal, show close matrilineal relatedness, based on DNA, when found in smaller groups, raising the question of whether data collected from large aggregations can readily help discern fundamental social units.

Below Bottlenose dolphin engaging in socializing behavior, which reinforces the bonds within a group at Chanonry Point, Moray Firth, Scotland, one of the most northern populations of bottlenose in the world.

Fission-fusion dynamics

Here is a typical and simple 30-minute scenario, but these interactions can be considerably more complex. First 3 minutes: Crooked Fin (mother) and Cookie (son, 7 month old) are alone (top left). Cookie leaves and joins his young male buddies, first Smokey and then Urchin joins (about 200 yards/180 meters from Crooked Fin). They play for about 9 minutes (right). Meanwhile, Crooked Fin forages but is then joined by her juvenile daughter (Puck), and they join Yogi (Smokey's mother). Smokey and Cookie join the group and Urchin goes off to join his own mother. Yogi and Smokey stay for 6 minutes then leave. Crooked Fin, Cookie, and Puck stay together, then Cookie leaves again, this time going off on his own and starts foraging.

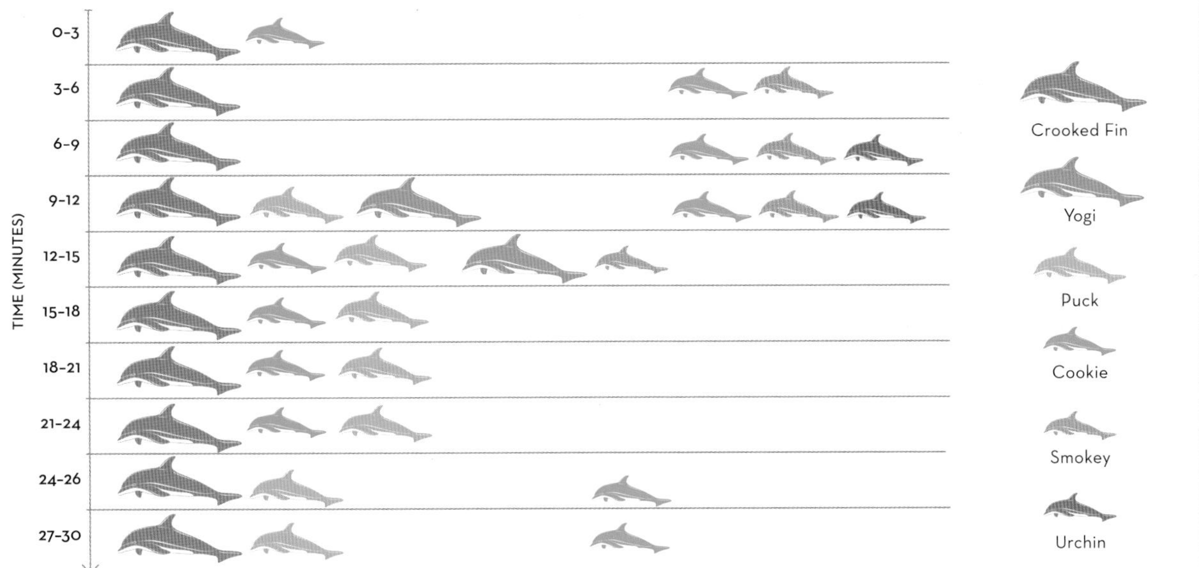

SOCIAL DYNAMICS

Although many think the term "pod" applies broadly to dolphin and whale species, it is a misnomer in most cases. Pod implies stability (such as peas in a pod) and group stability is not very common, although the core units of killer whale, pilot whale, and sperm whale society would qualify. "School" is applied to both fish and cetaceans, owing to the discourse of whalers and nautical men, but the term connotes anonymity—an arbitrary grouping of individuals. With variable terminology, many use terms such as unit (for a stable group), group (to mean any individuals that are associated), and community or subcommunity to refer to individuals that are at least loosely connected in a network. Most dolphin and whale species exhibit extreme fission-fusion dynamics, meaning that associations are temporary. Although some might associate for days at a time, many change who they associate with throughout the day.

These dynamics are key to identifying the function of social bonds. Those that spend more time together are more "bonded" generally than those that spend less time together. The definition of "together" or associated varies by species and by researcher. Some use spatial proximity. Others use a combination of spatial proximity and behavioral coordination, such as traveling in the same direction. Defining "associated" is difficult for pelagic delphinids such as common dolphins and Atlantic spotted dolphins that can number in the hundreds or thousands, and mysticetes, such as fin whales, who use low-frequency (< 100 Hz) communication that can travel over tens of miles. If social communication defines social units, unit boundaries would be impossible to measure for many species. Any definition of a social unit should be biologically meaningful. One useful rubric is to think of associates as those who are close enough to harm or help in the moment. With this guide, then spatial proximity is an appropriate measure. With fission-fusion, dolphins and whales can have context-specific associations based on common interests (raising young, finding food, defense, finding mates). This allows cetaceans to balance the costs (such as resource or mating competition and disease) and benefits of group living.

LIFE HISTORY OF CETACEANS

Life history concerns the timing and patterns of growth, survival, and reproduction of an organism. Animals that take a long time to reach reproductive age also live long lives and are said to have "slow" life histories. Life history parameters include gestation duration, physical size, brain size, growth rates, age of weaning, age of sexual maturity, age of last reproduction, maximum and expected lifespan. Cetaceans exhibit some of the most fascinating life history traits among mammals and these traits set the stage for every facet of social living.

LIFE EXPECTANCY

Cetaceans are at some of the extremes in terms of life history traits because many species live for decades. The bowhead whale is the longest-living mammal at more than 100 or even 200 years. Killer whales have also been documented as living for over 100 years. Wild bottlenose dolphins in Sarasota Bay, Florida, are living into their sixties. Sexual maturity (onset of reproduction) can be as late as the teens for many dolphin and whale species. Mysticetes have a faster life history strategy than odontocetes in that they grow faster, reach sexual maturity within a few years, and wean their offspring earlier. A prominent exception is the bowhead whale, where females reach sexual maturity at about 20 years of age and nurse their offspring for longer than most mysticetes.

DEVELOPMENT

Prolonged development, specifically a long infancy and juvenile period, is associated with social complexity and relatively large brain size. Life history theory does help explain this pattern, as it is a theory about trade-offs, namely: trade-offs between growth and reproduction, trade-offs between growth and survival, trade-offs between brain and body size, and trade-offs between survival and reproduction. Large animals, such as whales, tend have good survival and reproductive prospects, but it takes time to grow big. Animals that invest in growth must "wait" to reproduce because reproducing too early detracts energy from growth. Reproduction, and especially early reproduction, also increases mortality risk for both sexes because both are shifting effort and resources away from self-maintenance and toward whatever it takes to reproduce. And, finally, growing and maintaining a large brain is energetically expensive, more so than any other organ (see Chapter 2). Dolphins, compared to the larger mysticetes, exemplify the pattern whereby species with larger brains have slower growth and reach adult size later. Put simply, it is difficult to grow a big brain and a big body at the same time. Presumably natural selection has favored brain-size expansion in dolphins and body-size expansion in whales. There is good evidence for this in the anatomical record of ancient cetacean skulls and bones—which show the divergence of odontocetes and mysticetes (see Introduction, page 11).

With a prolonged developmental period, there is also a longer period of learning before facing the challenges of adulthood, namely reproduction. Given that it takes dolphins a decade or more to reach maturity, how they spend their time is critical. The most common hypothesis of a prolonged juvenile period is the "learning hypothesis," whereby delaying reproduction allows young to develop social, cultural, and ecological skills that favor later survival and reproductive success. This is effective only if adult and juvenile mortality is low—otherwise speedy reproduction would be the way to succeed. In species with high social complexity, larger brains would be favored, which would in turn favor slower growth. Those that spent their youth well would be better prepared to take on the vagaries and intricacies of social life. Primates, dolphins, elephants, and a few other species fit this pattern—even though the extent of social complexity for many delphinids is not yet known.

Opposite A pair of beluga whales in the White Sea, Karelia, Russia. Belugas can live for up to six decades, although global warming is threatening this Arctic species as the decrease in sea ice limits the amount of algae, the base of the food chain on which they depend.

Estimated lifespan & gestation period of cetaceans

Dolphins, whales, and porpoises vary widely in their life history traits. Because populations vary and we do not have enough long-term data to be certain, ranges are given for some of the best-known species.

SPECIES	AVERAGE LIFESPAN (IN WILD) IN YEARS	GESTATION PERIOD IN MONTHS
Beluga (white) whale	40–60	13
Blue whale	70–90	11
Bowhead whale	100–200	13–14
Dwarf sperm whale	20–25	9–11
Fin whale	60–100	12
Gray whale	50–70	13.5
Humpback whale	40–100	11
Minke whale	30–50	10
Narwhal	40–60	14
Pilot whale	45–60	12–16 (long-finned); 15 (short-finned)
Sperm whale	60–80	16–19
Amazon river dolphin	15–20	9
Baiji or Chinese river dolphin*	20–25	10–11
Bottlenose dolphin	25–60	12
Irrawaddy dolphin	25–30	9–14
Orca (killer whale)	50–100	15–18
Spinner dolphin	20	10
Striped dolphin	45–60	12
Harbor porpoise	25	10–11
Vaquita	20	10–11

* extinct since 2006?

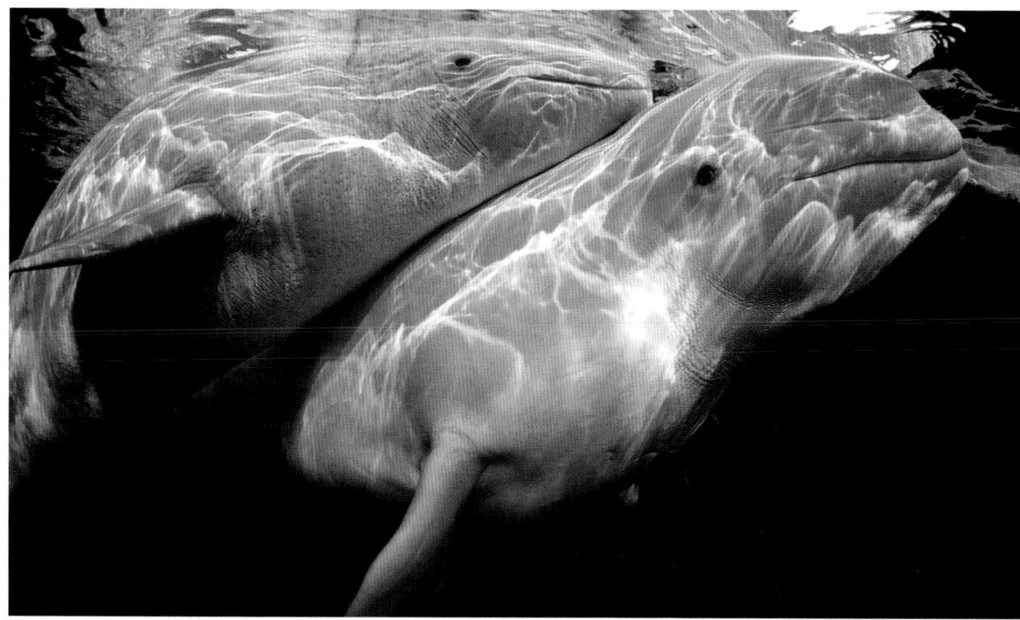

FAMILY BONDS & GRANDMOTHERS

In odontocetes such as killer whales, bottlenose dolphins, and pilot whales, a female might have several generations of offspring that associate with her. One adult female in our Shark Bay dolphin population is a great-grandmother, with five surviving offspring ranging in age from 30 to 6. Her 30-year old son checks in every once in a while, spending time with his siblings, nieces, and nephews. Persistence of family bonds across the lifespan could have important benefits. Kin have a stronger tendency to protect each other and share resources than non-kin. In other words, the benefits of group living are enhanced among kin because they tend to cooperate and share more than unrelated individuals. This is sometimes referred to as kin-altruism.

For killer whales, short-finned pilot whales, and possibly false killer whales, this is taken one step further, showing evidence of menopause, a trait once thought to be uniquely human. Menopause has consistently puzzled evolutionary biologists because why would natural selection favor cessation of reproduction? Some have posited that menopause is just a by-product of modern living where humans outlive their normal reproductive lifespan. Theories about menopause are varied and complex, but there is growing consensus that menopause is not really about ceasing reproduction, but about prolonging the lifespan. In all the species mentioned, females stop reproduction in their forties, but live until their sixties or later. Granny, the infamous killer whale in the waters off British Columbia, was born circa 1911 and was still swimming around with her elderly offspring until her death in 2016. As mentioned earlier, there is a trade-off between reproduction and survival. If survival is favored, it is difficult to favor reproduction simultaneously. All whales with menopause have their adult offspring still living with them in strong matrilineal units—where neither sex disperses from the group. The preferred hypothesis is that post-reproductive females provide benefits such as information and resources to adult offspring. If they reproduced until death, the last few offspring would be unlikely to survive, and perhaps the adult offspring wouldn't do very well either.

Evidence in support of this comes mostly from over four decades of research into fish-eating killer whales. The eldest matriarchs are the leaders of their groups, and this becomes more apparent when salmon supplies are low. Killer whale females share fish with male kin, who, with their larger size might be less able to maneuver to catch prey. In addition, killer whale adult sons experienced higher than expected mortality following the death of their mothers. Finally, short-finned pilot whales live into their sixties, unlike the closely related long-finned pilot whales, which do not have a post-reproductive lifespan, and die in their forties. It is curious as to why, with similar social structures, short-finned pilot whales found the secret to long life while long-finned pilot whales did not. Such mysterious patterns cry out for scientific investigation.

Left Granny, the infamous killer whale in the waters off British Columbia, evaded her would-be captors in the 1960s and 1970s when so many killer whales were being consigned to captivity in marine aquariums for human entertainment. She may have been as old as 105 when she finally went missing and was presumed dead in late 2016. Orca matriarchs play a crucial role in the group, often taking care of other females' offspring.

Opposite Both long-finned (top) and short-finned (bottom) pilot whales are highly sociable and remain in family groups, making them an easy target for hunters in Japan and Norway.

THE MOST BASIC SOCIAL UNIT: MOTHER & CALF

As in many mammals, the closest and most persistent bond in cetaceans is between a mother and her calf. But unlike in most mammals, the mother-offspring bond can last for decades—a lifetime. This intimate relationship provides a window into many facets of social life, including patterns of social behavior, how relationships are formed and maintained, the amount of tolerance shown for others, and the larger social network.

The relationship between mother and offspring is the key component in defining social structure. First, the mother-offspring unit is the most basic social unit. Second, the persistence of the mother-daughter or mother-son bond plays a large role in defining social organization. Most female mammals practice philopatry, that is, they remain with the social group or in the location of their birth for life. Because of this, most social groups are based on matrilineal, not patrilineal kinship. Paternal care has not been documented for any cetacean, but male relatives sometimes provide direct or indirect assistance in a few species. Among many cetaceans, the core unit is the matrilineal unit, comprised of a mother, her dependent offspring, female kin, and their dependent offspring.

NURSING & WEANING

Cetaceans invariably give birth to one offspring at a time. Although mysticetes and porpoises sometimes give birth annually, longer intervals of several years are more typical given the long gestation and lactation periods. Because of the enormous investment in the physical development of each calf, it is virtually impossible for a female to nurse twins, which are estimated to occur in less than 1 percent of cetacean births. Even in captivity, both twins have never survived. Lactation is extensive in cetaceans, but entails two basic strategies. One strategy, typical of mysticetes, is to produce prodigious amounts of milk that is very high in fat to support rapid growth. For example, Northern right whale calves gain about 75 lb (34 kg) per day in the first 12 months of life. Mysticete milk can be more than 50 percent fat (human or cow milk is 4-6 percent fat, and the richest ice-creams on the market are 25-30 percent fat), and mysticete mothers fast for the initial stages (or all) of lactation. Not surprisingly then, mysticetes nurse their offspring for less than a year and, once the calf is weaned, s/he has only brief encounters with kin. For most mysticetes, by the time mother and calf have reached the feeding grounds, their relationship has waned. Mother-offspring bonds are not sustained over the long haul and across vast oceanic migrations.

ODONTOCETE LEARNING & DEVELOPMENT

In contrast, odontocetes produce slightly lower fat milk (20-40 percent), but nurse their young for years. Mothers do not fast during lactation, but are known to feed infrequently during the first weeks of a calf's life. This is likely because the mother cannot dive to feed without leaving her vulnerable calf at the surface. As calves gain swimming, diving, and breath-holding skill, they can remain close to their foraging mothers. Weaning ages are not well known for most odontocetes for several reasons. First, unlike mysticetes that sever associations at the end of lactation, odontocete calves might continue to associate with their mothers after weaning, and sometimes, for life. Second, longitudinal study, which involves tracking the same individuals across years, is needed to document weaning age.

Weaning ages have also been estimated based on growth rates, vestiges of milk in stomachs, and chemical analysis of tooth growth layers, but mostly for dead animals whose exact ages are not known. Also, such methods are inadequate for detecting gradual weaning, which might be characteristic of many odontocete species. Fortunately, odontocete calves in various species (such as spotted dolphins, and killer, pilot, and sperm whales) swim in a characteristic position we call "baby position," under the mother's tail and abdomen—the position from which nursing occurs. While calves go in and out of baby position frequently while still dependent, and might catch their own prey too, the cessation of the baby position is a good indicator that the calf is weaned. In Shark Bay, bottlenose dolphin calves spend about 30 percent of their time in baby position and nurse periodically up until the time they are no longer "allowed" in baby position. Female mammary glands are visibly swollen while the mother is lactating, indicated by two bumps under the tail. Shark Bay bottlenose calves nurse on average for four years, but many are weaned at three years of age, and some as late as eight years. Weaning typically occurs when the mother is pregnant with the next calf.

This remarkable variation in weaning age has been documented indirectly in other odontocetes. Stomach contents of short-finned pilot whales captured in a Japanese fishery were examined in one study and found individuals as old as 15 years with milk in their stomachs. Typically, whales older than four did not have milk. Of course, in these instances, we don't know if the mother or another individual has provided the milk. Additionally, nursing has to have occurred within a couple of hours before death to be detected postmortem. That might be particularly unlikely under the stresses of a mass stranding event or when human whalers are in pursuit.

MYSTICETE LEARNING & DEVELOPMENT

Setting such debates aside, we do know that the period of dependency in odontocetes is intimate, extensive, and characterized by a long period of learning. Through the mother, calves learn to navigate a complex social and physical environment. Mysticetes have a different set of challenges because they travel from the relative safety of warm breeding waters, to the cold but productive high latitudes. Predation risk can be quite high during the migration, as killer whale attacks on humpback and gray whale calves have been well documented along the US and Australian coastlines. Once the calf is weaned, it must fend for itself entirely. Among odontocetes, the mother might play a protective role for years or even well after weaning if her offspring remains with her.

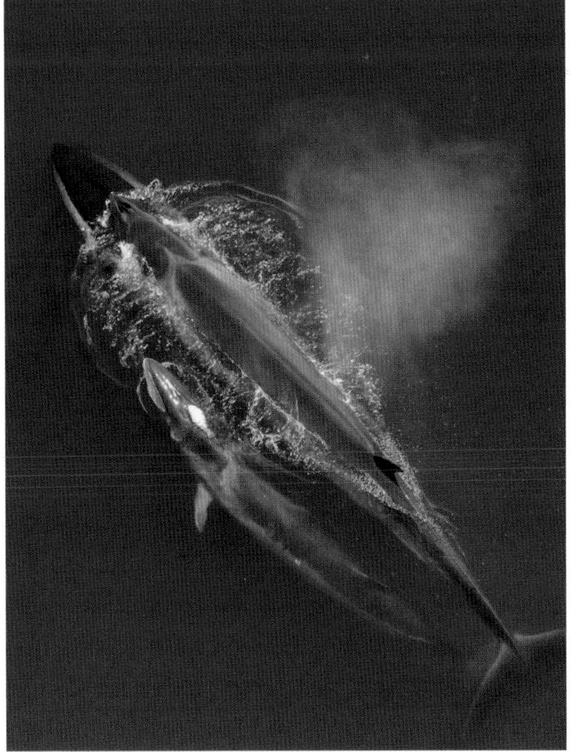

Above right White Southern right whale mother and calf. It is common for whale calves to swim above their mothers where they are protected from killer whale attacks. Calves nurse during the first year and are weaned before the mother begins her migration back to the breeding grounds.

Right Sei whale and calf. Very little is known about sei, fin, and blue whales, which are the largest species. Like humpback whales, it is likely that calves learn important migration routes and feeding techniques while following their mothers.

Above Odontocete calves swim in infant or baby position, under the mother's tail, lightly touching her abdomen. When breathing, the calf briefly moves alongside (top) before resuming baby position (bottom). The position allows the calf to get a hydrodynamic benefit, although, like echelon swimming (see page 112), at some energetic cost to the mother.

Data on social development in mysticetes are virtually absent. Outside of the mother–calf bond, there are few reports of calf social interactions beyond an occasional play bout. As indicated by their larger social organization, relationships are not well differentiated. In the breeding areas, mother–calf pairs actually avoid other mother–calf pairs (humpback whales and right whales)—completely opposite to what odontocetes do. The reasons for such avoidance are not well understood. Why wouldn't mothers want their calves to move about more and play with other calves? There are various hypotheses that try to explain this phenomenon. One is that mother–calf mutual recognition is low and until the calf can recognize the mother and vice versa, they might confuse offspring. Another related hypothesis is milk theft. Given the high fat content of milk and how many calories mothers provide in each nursing event, a calf that "steals" milk from another female might gain a significant benefit, but at a loss to the female's own calf. It might be difficult to prevent such theft other than avoiding mother–calf pairs altogether because of the amount of milk the females are producing. In point of fact, the rare occasions on which mysticete mothers and calves have been seen together, calves have been observed stealing or attempting to steal milk from other females. These females appear irritated and try to prevent the crime by swimming very fast, resting their belly on the seafloor, or turning belly-up. During the migration, when the calf is much older and bigger, mother–calf pairs are more tolerant of each other and sometimes travel together, possibly due to their mutual interest in reducing killer whale predation, although little is known about social behavior during the migration.

ODONTOCETES & NATAL ATTRACTION

In contrast, odontocetes are more sociable from the start. Calves are the center of attention. Among bottlenose dolphins, groups are larger during the newborn period for the first few months of life and in the first year of life than when the calf is older. Mother–calf pairs are not only mutually attracted to each other, but other females, particularly young females, find young calves irresistible, a phenomenon we have called "natal attraction." Newborn calves, with their lumpy look, distinct fetal stripes, cork-like or jerky movements, and large heads appear to fit the definition of "cute" that Konrad Lorenz, the famous Nobel-prize-winning ethologist, identified many decades ago. Juvenile females take this to an extreme by attempting to steal the newborns, if only for a moment, away from their protective mothers. The would-be bandits can accomplish this feat by swimming very fast by a newborn calf in the first week or so of its life. The calf will instinctively follow any fast-moving object (like many precocious young animals, including newly hatched goslings that Lorenz famously studied), and follows the thief until the mother chases them down and swims quickly alongside, taking her calf back. By the second week, the mother does not seem concerned about her calf's wanderings, and allows the newborn to play and swim with others. This suggests that there is mutual recognition by this phase, a pattern supported by whistle patterns studied in captivity.

CALF CARE & DEVELOPMENT

New mothers have a very high rate of signature whistles just before birth and soon after birth. It is likely that the young calf learns to recognize his or her mother this way by the first week of life—a process called imprinting. What do the juvenile females get out of it—other than the wrath of the mother? Well, it is thought that this attraction benefits females because they can gain parenting experience well before they have their own. One would predict that young females with such experience would be more successful in rearing their calves than females without such experience, something that would be hard to test for in the wild.

Mothers and calves are often found together and this allows the calf numerous social opportunities. Playgroups can be large and intense, with calves and sometimes juveniles joining in with leaping displays (see page 97), and various forms of body contact. Many of the behaviors observed during this period are likely critical for developing social skills, but also form the basis of the first social bonds outside of their relationship with the mother. These bonds can have long-term consequences. One such case is described in Chapter 1 for two bottlenose dolphin calves, Smokey and Cookie. Calves also form relationships with kin, matrilineal and sometimes patrilineal, although we know little about the mechanisms of kin recognition, other than by association. That is, older offspring would know who their younger siblings are and offspring would learn which individuals are part of their mother's network. In stable kin groups, the calf would easily learn who is who. It might be more challenging in dynamic fission–fusion societies involving hundreds or even thousands of individuals.

SOCIAL ENCOUNTERS

Dolphins, whales, and porpoises share the same problem: how to find each other. Terrestrial mammals can leave chemical traces or have territories or reliable ranging patterns they can use to find each other. Or there might be a common resource such as a waterhole that enables individuals to find each other reliably. Cetaceans can rarely use such cues, especially over vast distances. Sound, which travels so well in water, is probably the most dependable medium for communication and, unsurprisingly, what cetaceans commonly use. As you have seen from Chapter 4, the type of communication they use is tailored to the specific social problem the cetacean needs to solve, such as choosing a mate, finding your calf, or recognizing your kin or your best "friend."

The social lives of animals are largely defined by the nature and frequency of encounters that they have with others. Mysticetes are considered to be solitary and meet up during courtship and mating and might aggregate at large food patches. The extent to which mysticetes have individual recognition is not known. The same individuals sometimes engage together in bubble-net feeding (see Chapter 7) in both the Atlantic and Pacific oceans, but it is not known whether these repeat encounters are incidental or intentional, based on shared habitat on the feeding grounds.

RECOGNITION

In odontocetes, there is considerable evidence for individual and kin recognition based on the patterns of association and cooperation between individuals. Stable groups, such as sperm whales and killer whales, have calls specific to their matrilineal unit, called dialects (see chapters 4 and 6). They use these to maintain contact with their kin group. Pilot whales, which also have stable units, might also have such calls, but data here are preliminary. Long-term close kinship bonds are characteristics of these species. As might be expected, there are high levels of cooperation between members of these units. Sperm whales protect calves from killer whale attacks by forming a circle around the calf, sometimes with their heads facing out, sometimes with their heads in the center, so their powerful tails can defend the calf. Killer whales exhibit exquisite levels of cooperation when it comes to capturing marine mammal prey (such as wave-washing, see Chapter 7), and sharing food. In both sperm whales and killer whales, they can spread out to hunt, and use their unit-specific calls to both keep in contact and to reunite.

Among the smaller dolphin species characterized by extreme fission–fusion, individual recognition, rather than pod-specific dialects, seems to be important. Signature whistles are one mechanism for individual recognition, but this might not be the only way. Once in close proximity, they might use chemical and physiological features. Regardless of the mechanism, it is clear that relationships are important. Scientists measure these social bonds primarily by examining association. They use a coefficient of association, which is a way of calculating how much time animals spend together. Those that are always together would have a coefficient of 1.0. Those that are never together would have a coefficient of 0. Social bonds may then be characterized in several ways: avoidant relationships, casual acquaintances, and preferred associates. Avoidant relationships would be individuals that are rarely seen together despite overlapping home ranges. Casual acquaintances are those that interact periodically, but not much more than one would expect given their shared home range. Preferred associates are the closest bonds and these are individuals that are seen together much more than expected at random.

Opposite top A young bottlenose dolphin calf in the Red Sea, Egypt rubs on his mother's head and melon. When not in baby position, young calves spend quite a bit of time near the mother's head, possibly listening to the sounds she produces.

Opposite bottom Two adult bottlenose dolphins in Shark Bay, Australia, swimming synchronously, side by side. Synchronous swimming indicates a close social relationship.

PROXIMITY

Spatial proximity is a good indicator of the strength of a relationship. Even within stable groups, one can examine who is swimming together closely and who is on opposite ends of the group. Behavioral biologists sometimes collect "nearest neighbor" data at regular intervals as a measure of who an individual's closest associate or "friend" is. Scientists have also used photographs of dolphins and whales to examine who is in the same frame as a measure of association.

Cetaceans that swim together often breathe together, called synchronous breathing. Synchronous breathing can be so well timed that individuals are actually exchanging respiratory droplets, akin to kissing or coughing on one another depending on how you think about it. Synchronous breathing rates are high for closely bonded individuals, such as a mother and calf or members of a male alliance in bottlenose dolphins. Synchronous breathing is not without risk, as it is thought that some deadly diseases, such as morbillivirus, a disease that kills thousands of dolphins each year, is transmitted through respiratory droplet exchange.

CONTACT BEHAVIORS

Contact behaviors are also important demonstrations of either affiliation or aggression, depending on the nature of the contact. Many dolphin and whale species pet each other (using their pectoral fins or flukes), and rub each other (body to body contact), in what most would describe as affectionate contact. The mother-offspring unit can provide important clues on affiliative behavior because they tend to engage in synchrony and contact behaviors at high rates.

Contact swimming is common among cetacean species and can take various forms. Echelon swimming is widespread between a cetacean mother and her young calf and involves swimming side by side either parallel or with the calf slightly ahead, often with the calf's pectoral fin resting on the mother. This type of swimming provides a hydrodynamic boost to the calf as it is pulled in her slipstream, but at some energetic cost to the mother. Sometimes older animals swim in this position, although the context is not well known. Similar contact swimming behaviors include flipper to flipper contact, akin to holding hands. To stay in this position takes close coordinated swimming. Another form of contact swimming is what we call "bonding." This behavior, which we have witnessed extensively in Shark Bay bottlenose dolphins, involves one individual, the bonder, placing their pectoral fin on the flank of the other, but near the tail such that the two are staggered.

What is fascinating about this behavior is that it occurs predominantly between females in the context of male harassment. Bonding takes on a dramatic form with this typical sequence. It begins with an adult female who is being consorted by two or more males. Such males are often being aggressive by chasing, popping (see Chapter 4), pushing up the female, and displaying around her. Another female joins the group and quickly swims over to the female placing her pectoral fin on her flank. The females might remain glued to each other for 30 minutes or more. Males do not seem to appreciate this activity and will sometimes attack the bonder or swim between them to break them up. Importantly, males sometimes stop their attack, fall behind, and/or slink away. On rare occasions, two juvenile males might bond and—even rarer—an adult female and male might bond, but not when she is cycling and able to get pregnant.

For young dolphins and whales, social play is a common way to test one's skills and to develop social bonds. Like terrestrial mammals, cetacean play has many similar characteristics in that it appears to be "mock" versions of adult behavior. For example, bottlenose dolphin play often involves chasing, courtship displays, and mock fighting. Just as adult males engage in synchronized displays behind or around a female, so do male calves and juveniles, usually behind a young female, but occasionally around another young male.

PETTING

Perhaps the most intimate form of affiliation is expressed with petting and gentle rubbing. This can be expressed mutually or be one-sided with one individual petting another. Just like primate grooming or bird preening, these behaviors probably began with a hygienic function, removing skin parasites, and became a predominantly social behavior. In grooming, preening, and petting, the receiver is likely to present preferred body parts to be touched, and these are often sensitive areas, such as around the blowhole, or near the eye or the genital area. Petting and rubbing events are usually brief, lasting a few minutes, but can sometimes occur for much longer. Mother and calf commonly pet and rub each other, but so do male members of an alliance and females that have been friends for a long time. Male-female petting also occurs in the context of mating, and might indicate some element of female choice. This type of contact is widespread in cetaceans, and requires further study.

Why are contact behaviors so important in social life? Why are they key indicators of close social bonds? There is a fascinating literature on this with some major insights developed by an Israeli biologist, Amos Zahavi. He developed an idea based on a handicap principle. To put it simply, Zahavi proposed that honest signals must carry a cost in order to be "believed" by the receiver. Otherwise, cheating or deception would be too easy. Hence, the expression about human deception where "talk is cheap." One way to incur a cost is to handicap oneself or make oneself vulnerable to the other. Close contact necessarily makes each party vulnerable to the other. Grooming in primates, petting and rubbing in cetaceans, and many other rituals of affiliation, including handshaking, kissing, hugging, stroking, and patting in humans, provide fairly accurate data on the status of the relationship.

Opposite (top) Google, a young Shark Bay male bottlenose dolphin sponger (see Chapter 7), petting his adult sister, Gimp, and (bottom) two young females engaged in contact swimming. Sometimes dolphins swim pectoral fin to pectoral fin—akin to "holding hands" or just with a pectoral fin resting on the side of the other. If a dolphin rests his or her pectoral fin on the flank of the other, but is slightly behind (not parallel, as shown here), we call it bonding. Bonding occurs almost exclusively between females and is in the context of male harassment.

FRIENDS, ALLIES & AGGRESSIVE ENCOUNTERS

Concepts such as "friend" and "ally" may seem anthropomorphic, but biologists have found these terms useful to characterize bonds in many social species such as primates, hyenas, meerkats, horses, lions, and cetaceans. Simply, friendship can be defined as when two individuals reliably and preferentially associate, show mutual tolerance, and exchange benefits, something scientists call reciprocal altruism (expressed by the adage "I scratch your back and you scratch mine"). Friends might groom each other, share food, help raise each other's offspring, and assist each other against a third party. As such, friends can also be allies. Allies or alliances are less defined by friendly behavior or the exchange of benefits within the pair, but more by routine cooperation against third parties. Friends can be allies and allies can be friends, but one is characterized more by aggression toward others and the other more by affiliation within the relationship.

As described in the Holeyfin-Hobbit episode on page 94, Surprise and Holeyfin could be considered friends because they were affiliative over a long period of time (more than ten years and until Holeyfin's death) and both supported and defended each other against sharks and other dolphins. The term "friend" is generally not used to describe relations between kin. This is in large part because the benefits of helping kin are obvious, as kin share genes by descent. Even though kin groups are common in odontocetes and close bonds between kin are characterized by support, defense, and even food-sharing, the term "friend" is reserved for unrelated individuals. As described earlier, relationships between unrelated individuals can be described as avoidance, acquaintance, and/or preferred. Preferred associates might also meet the definition of "friend," depending on what criteria one uses.

COALITIONS

If the exchange of support is short term, opportunistic, and involves active defense, it is referred to as a coalition. Coalitions are usually restricted to specific circumstances in which two or more individuals cooperate against another individual or group. For example, several females might form a coalition to chase off an aggressive male because it benefits all females to do so. Coalitions can be reciprocal from one day to the next, but are distinctly short term. Coalitionary partners can also be "friends" if the bond is durable but mutual agonistic aid is temporary. Friends are more likely to rapidly form coalitions if conditions call for it. If the same coalition repeatedly acts jointly against others, then the bond is durable, and they are not only friends, but allies.

ALLIANCES

The term alliance refers to repeated collaboration or cooperation with preferred partners within a community against other parties either within or external to a community. Like coalitions, alliances are borne out of direct contest competition with others. Friends can be allies or form coalitions, but the stability of the bond and its impact on others over time is what determines the type of relationship. Admittedly, the nature of the relationship can be difficult to determine given that some aspects are revealed only under challenge or over a period of time.

Because alliances are based on agonistic competition, male alliances are typically grounded in mating competition and female alliances tend to form over food/resource defense. There are certain

Opposite Two subgroups of juvenile male bottlenose dolphins clustered by the relative strength of their bond. Juvenile males spend considerable amounts of time together socializing and apparently trying to figure out who their adult alliance partners will be.

Bottlenose dolphin male alliances

The social network and clusters of the Shark Bay bottlenose dolphins. While the entire network shown here is 439 individuals, shown by the circles, they are in six clusters, identified by the color of the circles, reflecting who associates with each other the most.

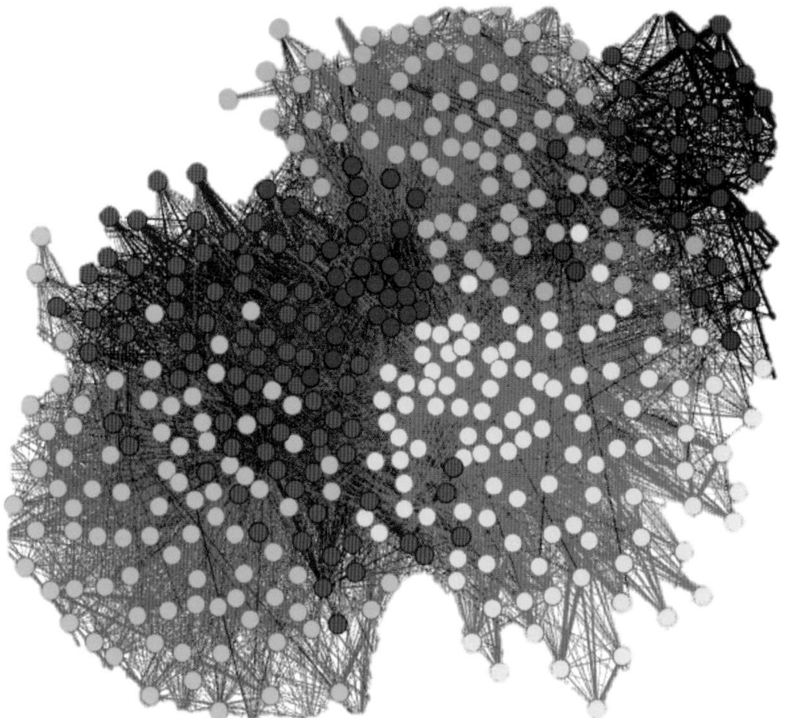

Light blue and purple clusters contain all the sponge-using dolphins described in Chapter 7, in addition to many non-tool-using dolphins. All clusters are comprised of both males and females, but the red cluster is all-male, with one exception. The thickness of lines connecting the circles (individuals) shows the strength of their connections.

advantages to forming alliances with kin because the spoils can either be shared or, if not shared, at least kin receive the benefits. This becomes acutely obvious when we compare alliances across the animal kingdom. Kin-based female alliances are common in spotted hyenas and many Old World primates, such as rhesus macaques and savanna baboons, which fight over food patches, or in the case of spotted hyenas, freshly killed carcasses. In groups with female kin-based alliances, there are often strict female hierarchies based on matrilineal membership.

Males form kin-based alliances in which they either remain in their natal group, or leave their natal group together. Male lions prefer to form alliances with members of their natal pride, which is comprised of closely related females. Unlike the depiction in the *Lion King*, male kin cooperate to take over prides of unrelated females (not the one they were born into). The size of the alliance is important in determining their ability to secure access to a female pride. Male chimpanzees also tend to form alliances with close kin, which is easy to do since they remain in the community in which they were born. For both lions and chimpanzees, male alliances are all about enhancing mating opportunities. The same applies for bottlenose dolphin alliances, which are typically not kin-based, but are sometimes comprised of males that are more related than would be expected by chance (that is, random pairings in the population).

Although there is considerable interest in the social intelligence underpinning the ability to form and maintain alliances, they are extremely rare in the animal kingdom. This is because if males cooperate to enhance mating success, fertilizations cannot be shared. If two males cooperate to secure access to a female, only one can be the father. This creates a disincentive for males to cooperate. And, if one male can dominate the matings, why should he share? In some species, such as humans and hamadryas baboons, males cooperate to defend their respective family units. In other words, they largely respect the pair-bonds of other males, but benefit by defending their community against another raiding community. This occurs when boundaries between social groups are clear. In bottlenose dolphins, there are no pair-bonds to defend and community boundaries are not well defined. Thus, male alliances benefit by outcompeting other alliances, sharing in mating and possibly distributing mating success (fertilizations) among members of the alliance in sequential consortship events. Temporary coalitions would not be particularly effective because in dynamic fission-fusion social systems, a male might not have his partner around at the right time. Thus, bottlenose dolphin alliances tend to stay together or at least near each other, all the time. You never know when you might be challenged by another alliance—or find a female—or perhaps you cannot trust your partners when they are out of sight!

BOTTLENOSE DOLPHIN ALLIANCES

Bottlenose dolphin male alliances have been well studied for three decades now by Richard Connor in Shark Bay, Australia. Some of those male bonds have endured that long. Alliances are identifiable by consistent association between two or more males where they spend more than 50 percent of their time together. This is called a first-order alliance. Some males are in pairs, trios, or larger alliances. Besides association, members of an alliance engage in a suite of behaviors that demonstrate the importance and benefits of their bond. First, males swim tightly together and in synchrony. They often pet, rub, and mount each other. They usually separate a bit to forage, but come back together to rest, travel, or socialize. Second, the alliance cooperates in securing a single female during the breeding season. Sometimes the males rush upon the female and sometimes they attack her, especially if she tries to escape. Occasionally males simply join up with a female and swim behind her without overt aggression. Admittedly, this might be intimidating enough for a lone female. The alliance and the female form what is called a consortship, which can last for hours or—rarely—months, where the males swim alongside or behind the female in formation and continue to engage in synchronous surfacings and displays. Males even synchronously mount the female, effectively sandwiching her between them—but synchronous mounting shows the tolerance the males have for each other in the mating context. An alliance is sometimes challenged by another alliance that is trying to get access to the same female. This can result in battle, where members of one alliance try to defeat the other and end up with the female.

Below Two adult male lions in an alliance. Like dolphins, the males move in synchrony.

Since numbers matter, an alliance might also recruit their "buddy" alliance, in what is called a second-order alliance, to assist in the battle. Thus, the fight can involve four alliances or more—all over a single female. Presumably, the assisting alliance might be helped to secure a female on a different day. You might well ask why there is such intense competition over a single female? As mentioned above, nursing can last for three to eight years in Shark Bay. Since females become pregnant in the last year they are nursing a calf, a successful female is usually fertile every four years or so. That means there are many fertile males swimming around the bay but very few fertile females.

Depending on how one defines an alliance—by behavior, degree of association, or some combination—bottlenose dolphin populations differ in this trait and provide some insights into why alliances might form. Besides Shark Bay, bottlenose dolphin first-order alliances have been observed in Port Stephens, Australia, Sarasota Bay and possibly other sites around Florida, and in the Bahamas, but not in the Moray Firth of Scotland or Doubtful Sound, New Zealand. Second-order alliances have not been definitively reported at other sites. Why alliance formation occurs in some populations and not others is not entirely clear, but one clue might lie in relative body size differences between the sexes, sexual dimorphism. Males and females are very close in adult body size at the locations where males form alliances, but males are much larger than females in the Moray Firth and Doubtful Sound. Conceivably, relaxed selection on body size in colder temperatures might allow males to become large enough that cooperation with other males to secure mates is not necessary. That is, a single male can achieve mating success on his own. In the smaller bottlenose dolphins, lone males cannot dominate individual females—they are all the same size—and, once alliances occur, it becomes evolutionarily stable in that males who are not in alliances cannot effectively compete. Thus, the dominant strategy becomes alliance formation. Notably, in Sarasota, males are a bit larger than females and here males consort with females singly and in pairs. In Shark Bay where there is little sexual dimorphism and the bottlenose dolphins are the smallest found in the world, alliance size ranges from 2 to 14 animals. The high density of dolphins in Shark Bay might also mean that there is a high encounter rate with other males, further favoring alliance size because large alliances can generally outcompete smaller ones.

AGGRESSION

What does aggressive behavior look like in cetaceans? In bottlenose dolphins it is not uncommon to jaw other dolphins, ram them with their beaks, or whack them with their powerful tails. Juvenile and adult male bottlenose dolphins have lots of tooth rake scars, but also cycling females have fresh wounds from their encounters with males. In Doubtful Sound, New Zealand, adult males will butt heads much like rams do. Risso's dolphin males are often completely scarred up from fights with other males The narwhal male has a formidable tusk (actually a tooth), which is used for jousting. Male beaked whales have teeth and tusks that they use during battle with other males.

Jim Dines, the mammals collection manager of the Natural History Museum of Los Angeles, and colleagues did a review of all 89 cetacean species and found that 58 had sex differences in physical traits including in size, shape, teeth, and tusks. Most are related to male-male battle, but some, such as the large dorsal fins of male killer whales, might have been selected for because of female choice. The tall dorsal fin seems to be a health and age indicator. When killer whales encounter each other for mating, the female might not have much time to assess who is the best male to mate with and his dorsal fin might be an excellent cue. We know very little about female choice in cetacean mating systems. But most of the sex differences concern weaponry, such as tusks, teeth, and a robust frame for fighting or intimidating opponents. The enormous size of the male sperm whale, and particularly the head size, allows him to use his loud clicks (which are a good size indicator) to deter rivals from challenging him. It also provides an excellent cue to females as to which male is the biggest. Features that deter males from challenging each other might also be attractive to females when it comes to mating. As mentioned above, since males provide no paternal care, her choice might be based on whatever physical and behavioral traits are indicative of good genes for her offspring.

Opposite top Risso's dolphins are known for their extensive battle scars which accumulate with age. Since the scars are so distinctive and remain for a long time—rather than fade as they do in other species—it is thought that these indicate something about social status.

Opposite bottom Young males cooperate to sandwich a third between them. The "sandwiched" male is trying to get out of the way by changing direction.

CLIQUES, CLUSTERS & NETWORKS

What is a society? How is it defined? This is not an easy question to answer for any animal system, let alone human or cetacean societies. While some cetacean communities seem to have discrete boundaries, many cetacean species have overlapping networks that extend over hundreds, perhaps thousands, of miles. Some populations of cetaceans are defined by an ocean basin and others by a small bay or river. An important feature of many cetacean societies is that there are few physical barriers that might fragment a terrestrial population. But this does not mean they are not separated by land masses, ice, low tides, or other habitat features (temperature, depth, substrate characteristics) that affect the movements of and hence association between individuals. A society is usually defined as a group of individuals that associate or have a high likelihood of doing so based on geography.

Besides humans, few animal species have societies that number in the hundreds or perhaps thousands with individual recognition. We know virtually nothing about the limits of individual recognition in cetacean species, but through long-term studies we infer that dolphins and whales know each other based on behavior (frequent association, approaches, friendly interactions, signature whistles, or dialects). Over time, this information tells us something about the nature of cetacean societies and their structure.

A critical tool for characterizing cetacean societies is social network analysis because it allows for multiple levels of description, from the individual, to subgroups, to the entire population. While most of us are somewhat familiar with social networks through social media (Facebook, Twitter, Instagram, etc.), social network analysis uses math and fairly rigorous computations to quantify those connections. Typically, the individual (or other unit) is a "node" and the links or connections between individuals are called "edges." The individual or node can have multiple attributes, such as age, sex, reproductive status, or behavioral trait, and the edge can also have attributes, such as association, geographic location, or type of social interaction or relationship (kin, non-kin). So networks can have many features and

Above A large pod of bottlenose dolphins swimming along the Wild Coast of South Africa's East Cape, just south of Port St. Johns. The image was taken from the air during the annual sardine run.

Sperm whale social networks

Depiction of the three nested levels in the sperm whale society off the Galapagos Islands showing individuals within social units within vocal clans. Courtesy Cantor et al. (2015)

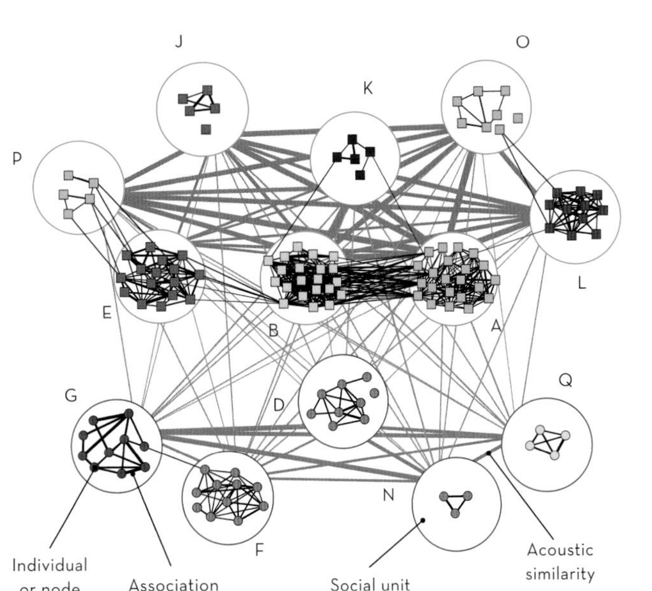

Regular clans, characterized by codas with regularly spaced clicks

Plus-One clans, characterized by codas with an extended pause before the final click

Social relationships (line thickness proportional to the time individuals were identified in the same group)

Acoustic similarity (line thickness proportional to the multivariate similarity of coda repertoires

Within the network, modules of individual whales (called "nodes" and shown as colored dots and squares) connected by their social relationships (black lines) define the social units (letter-labeled). In the overlapped acoustic network, modules of social units connected by similarity in acoustic behavior (gray lines) represent the vocal clans. The dialects are different between clans. Females who are members of different clans (and different dialects) might overlap spatially, but do not associate.

can be hierarchically structured (mother-calf, kin group, alliances, all associates). Significantly, it is not just the connections between individuals that is important, but one can also look at how an individual is positioned in the network. For example, do post-reproductive females have a central position in the network? This might suggest that they are important sources of information and connection between members of the society. Individuals can have many weak connections—"social butterflies"—or a few tight bonds, such as in alliances and strong cliques. In a clique, the members of the clique are all tightly connected to each other.

Societies can be hierarchically structured. Sperm and killer whales have highly structured social groups that are connected by varying degrees of matrilineal relatedness. At the society level, they form distinct cultures (see Chapter 6) based on behavior, dialect, or other attributes. Risso's dolphins have some structure in that subgroups are commonly found together, but there might be considerable fluidity between groups. At the other extreme are the many delphinid species that have high fission-fusion dynamics, changing groups over the course of minutes. Even so, all societies have some structure and interactions are far from random. One feature that most cetacean societies share is that they are sex segregated—in that females preferentially associate with other females and males preferentially associate with other males. The only real exception is those species that associate with opposite-sex kin. Sex segregation is common across many mammalian societies and is driven by a number of factors, some as simple as habitat or geographic preferences—safe areas for calves or where lactating females can support their offspring. But also, sex segregation can be driven by social factors—such as avoiding aggressive males.

The structure, including temporal, spatial, and social dynamics of cetacean societies have important implications for management and conservation (the subject of Chapter 8). The Hawaiian islands are a critical breeding area for humpback whales. Sound pollution from shipping, mineral exploration, or naval exercises can interfere with the communication that is necessary for finding others. Habitat destruction can wipe out a critical feeding or breeding area. Changes in water temperature and habitats can exacerbate stress and disease and impact network structure. For example, important individuals in a network might be most vulnerable to disease because they have high contact with others in the network. Understanding cetacean societies is not only of intrinsic interest—it highlights why they have some of the most complex social systems known outside humans—it helps us understand our potential impact on their survival.

6 DEEP CULTURE

Hal Whitehead & Luke Rendell

Untitled, Humpback Whale Mother and Calf Composite Photo
Kingdom of Tonga, 2006

"This is my first multi-image composite photograph of a whale. For twenty minutes the mother would circle me with her calf, coming closer with each pass. During her final approach, she passed three meters beneath me as her calf ascended to meet my gaze. In that moment, I composed my first life-size portrait of a whale."—Bryant Austin

HUMAN CULTURE, CETACEAN CULTURE

Life happens and creatures evolve because information is transferred. DNA is the main way that biological information passes down the generations. In animals that can learn from others, however, information can be passed from brain to brain as culture. This is especially true in our own species, but evidence is accumulating that culture is vital to the lives of several species of whales and dolphins. Songs and calls, feeding methods and migration routes, ecological niches and social networks are all aspects of cetacean life that individuals learn from others as they acquire the cultural knowledge vital for their survival.

HUMAN CULTURE

We humans are born with genetic templates handed down from our parents, but we cannot become fully human without what we learn from each other. As we grow, we benefit from a body of knowledge, skills, customs, and materials that always surrounds us. During our lives, we may add to, or modify this body, and future generations can build upon these changes. This is human culture, a large part of what makes humans human. To even visit the watery world of whales and dolphins, we have to use the seafaring knowledge and technology that humans have built up over many generations. The development started before 5,000 BCE when people first modified logs with some piece of material for the wind to catch. Today's fiberglass hulls and stainless-steel fittings are the cultural descendants of those logs—products of a system of cumulative cultural evolution that allows humans to cross oceans reliably, a remarkable achievement for a terrestrial mammal. The abilities to meld cultures and modify them before passing them on allows for the rapid evolution of extraordinary cultural products: jumbo-jets and the internet, hip-hop, and nouveau cuisine. The result has been the extraordinary evolutionary and ecological trajectory of human society over the past 10,000 to 20,000 years. We are clearly the only species on Earth to meld, modify, and exploit culture like this, but in the significance of culture in our lives, to what extent are we truly so different from all other species?

CULTURE BEYOND HUMANS

To some, the very idea of culture in a nonhuman animal is alien. Edward Tylor, the founder of cultural anthropology, famously described culture as "that complex whole which includes knowledge, belief, art, morals, law, custom, and any other capabilities and habits acquired by man [sic] as a member of society." From this perspective, culture is, by definition, uniquely human. More recently, however, as we have begun to understand animal behavior more deeply, many scientists have come to accept that some nonhuman societies have something that, while not the same as human culture, is nonetheless similar enough to merit consideration as being a form of culture. Such

Left A member of the Nez Perce tribe of the American Pacific Northwest, maneuvring a dugout canoe made from a large hollowed-out tree. These canoes suited the semi-nomadic fishing and hunting way of life of the Nez Perce. The process of development from these craft to modern ocean-going vessels is one of cumulative cultural evolution the accumulation and transmission of cumulative design improvements across generations), not genetic evolution.

researchers typically adopt a broad definition of culture, as we do in our book *The Cultural Lives of Whales and Dolphins* where we state: "Culture is information or behavior—shared by a community—which is acquired from conspecifics through some form of social learning."

Such definitions correspond roughly with the nonscientific concept of general culture as "the way we do things." Culture, then, is behavior or information with two primary attributes: it is socially learned and it is shared within a social community. Fundamentally, culture contrasts with genetic inheritance as a way that information moves from animal to animal. What is called "the question of animal culture" is still hotly debated in some academic circles, but the role of cultural processes in nonhumans is being increasingly appreciated, and perhaps nowhere more so than in the lives of cetaceans.

CETACEAN CULTURE

In the 1960s people started to study whales and dolphins in the wild, spending significant time observing their behavior. Prominent among these pioneers was acclaimed American zoologist Ken Norris, who spent a good part of his life with whales and dolphins, both wild and captive, observing carefully. After two decades of insights, in 1980 Norris wrote that the learning abilities of dolphins "translated into local variations in group behavior that we might call culture." By 1988 he had concluded that some of the social patterns that he observed were "clearly cultural."

Now we are beginning to fully appreciate the scope of cultural processes in cetacean behavior. A humpback whale learns the song he sings from other humpbacks, and in turn is a model for other singers. Over thousands of miles of ocean, the song culture of the humpback whale dominates the acoustic environment, as it has for millions of years. Young killer whales learn the specific and highly honed techniques that their group uses to obtain food and this learned behavior feeds back to affect their biological evolution over many generations, so, for example, the genes involved in digestion evolve along different lines in mammal-eaters compared to fish-eaters. Bottlenose dolphins open up new niches by inventing novel methods of catching fish, and form social networks built around these feeding lifestyles.

The "comparative method" in evolutionary biology is a way to try to understand how particular traits have evolved by comparing the presence and precise form of that trait across many different species. Where species in different lineages evolve similar traits, for example, this implies a similarity in the selection pressures that has led to convergent evolution. The parallels and contrasts with human intelligence therefore tell us much about cetaceans, as well as humans, and contrasting the social behavior of apes and dolphins can help us understand the forces behind the evolution of our own societies as well as theirs. In the same way, because some whales and dolphins have a form of culture that has evolved in a very different environment to our own, the fact that those cultures show some remarkable similarities to human culture, but also some profound differences, has the potential to reveal a lot about the evolution of culture in general. The cultures of at least some species of whale and dolphin, we have come to believe, are a major part of what they are.

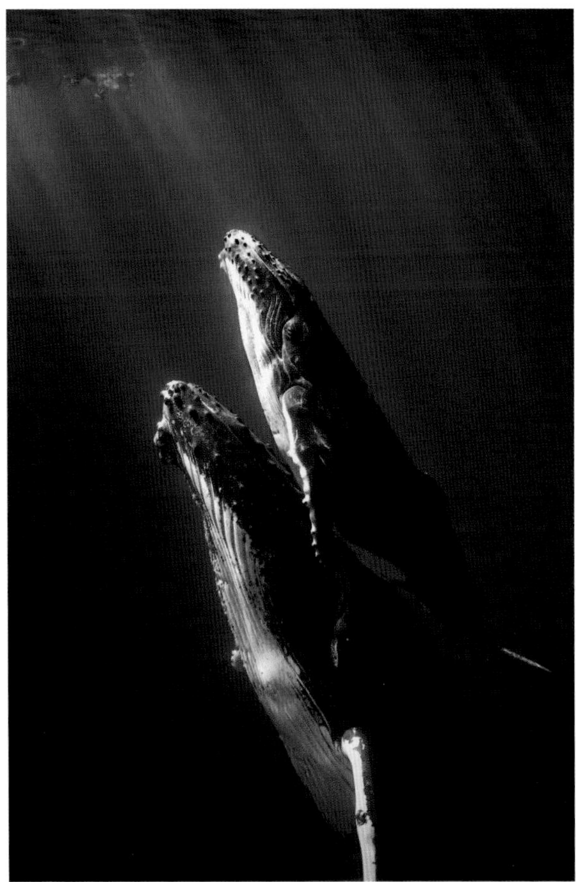

Above Young humpback whales have much to learn before they can become fully independent from their mothers. Calves are weaned after 6–10 months and reach sexual maturity between 4 and 10 years.

DOLPHIN LIFESTYLES: THE ROLE OF CULTURE IN SMALL CETACEANS

Among cetaceans, we are most familiar with bottlenose dolphins because they have colonized many varied ecological niches in the coastal waters nearest to humans, from deep fjords, to shallow, sandy bays. In coastal waters, their societies are organized around "communities" that inhabit particular geographic features or areas. The knowledge of how to make a living—meaning the foraging techniques a dolphin employs—in a particular place is a key part of dolphin culture. Where there are many ways to do so, populations can be structured according to which specializations an individual learns from its mother.

CULTURAL TRANSMISSION OF DIVERSE FORAGING STRATEGIES

Dolphins mostly eat fish, but not all fish are alike. Bottlenose dolphins have a surprisingly wide range of ways of making a living. They grub in sandy seabeds to unearth hidden prey. They slam their tails on the water surface to startle fish out of sea-grass beds. They use sponges as beak guards when foraging for prey hidden in rough seabeds (see Chapter 7). They use a "hydroplaning" technique to catch fish in water so shallow that it only just covers their pectoral fins. They use elaborate action sequences to catch, kill, and prepare cuttlefish for consumption. They can show division of labor in cooperative groups, for example, in "carousel feeding" where they work together to bunch prey against the surface, or in mud-ring feeding (see page 152), where one animal swims in a circle agitating bottom sediment to create a mud "net" that traps fish for their companions. They act collaboratively to create bow waves that wash fish up onto muddy shores, from where they are picked off by the dolphins that hurl themselves up the shore after them. We are a long way from mapping the true diversity of ways in which dolphins make their living, but we do know that no dolphin does all these things. The knowledge and tricks used by members of any particular community are distinctive and culturally inherited, mostly through the mother. This cultural diversity is how we think they have been able to colonize so many coastal habitats. It is their culturally driven success that has made them so familiar to us.

FORAGING STRATEGIES AS LIFESTYLES

In any one place, if opportunities are sufficiently varied, dolphins will find ways to specialize into distinctive foraging niches, and these "lifestyle choices" have important effects on the community. Janet Mann has shown how dolphin society in Shark Bay, Western Australia is structured around adherence to particular foraging specializations (such as those described in Chapter 7). In human societies, the way we make our living affects more than just the amount of money in our pockets—it guides our social interactions. Similarly, the way an individual dolphin makes a living, meaning the foraging tactics it employs, also has implications. The demands of particular foraging specializations can lead dolphins to associate more with others that share their tactics. This can lead to different approaches to socializing, longer hours, and different neighborhoods.

DOLPHIN CULTURES INTERACT WITH OUR OWN

Sometimes, humans create opportunities for dolphins, and the dolphins exploit them. A most extraordinary example is in Laguna, Brazil, where dolphins and human fishermen have been fishing cooperatively for many generations of each species. The origin of the cooperation is lost to living memory—it has been going on for longer than any of the humans or dolphins now taking part has been alive. The fishermen wade out with their hand-cast nets and make a commotion by slapping their hands and nets onto the surface water.

In response the dolphins herd fish while the fishermen watch closely for a distinctive type of arch-backed dive, the cue to throw their nets in front of the dolphins to catch the herded shoal. Scientists have confirmed that only a subgroup of the Laguna dolphins work with the fishermen (around half of a population numbering at least 60), and have shown that the dolphin community is socially divided between those that do and those that do not.

FROM SYNCHRONY TO CULTURE

Synchrony of behavior plays an important role in bottlenose dolphin society. Males that belong to "alliances" (coalitions that work together for access to females) often move with exceptional synchrony, to within tenths of a second. They perform coordinated, dance-like sequences of dives and turns around females. Scientists suggest this is a signal of the strength of the alliance bond. If stronger alliances mean more mating opportunities and more offspring, then dolphins may well have evolved a psychology that both gives them the ability to copy others accurately and also "rewards" them for achieving that ability. This could explain why trained dolphins are so adept at copying each other, and even humans, on command. The ability to copy is also key to understanding how dolphins are able to learn feeding techniques from each other. Thus the evolutionary struggle to pass on genes may have helped generate the cultural diversity of dolphin lifestyles we see today.

Top Some dolphins have mastered the technique of hydroplaning, pumping their tails to gain speed then using the momentum to carry them into the shallows to catch fish that would otherwise remain out of reach. This adult female named Reggae in Shark Bay, Australia is particularly adept at hydroplaning and has passed on the skill to her offspring.

Above A fisherman in Laguna, Brazil casts his net, responding to a dolphin's characteristic "humped" dive, a signal that it has herded a shoal of fish. The innovation and subsequent transmission of new feeding methods from mother to calf has resulted in an extraordinary diversity of ways in which bottlenose dolphins make a living, and more remain to be discovered.

BIG WHALE CULTURE: SONGS, MIGRATIONS & INNOVATIONS IN MYSTICETES

One of the most iconic pieces of whale behavior, the song of the humpback whale, turns out also to be a fantastic example of whale culture. Songs are complex and hierarchical in structure, but the most fascinating thing is how they change. All the whales in an area change their songs together, usually gradually, but in the South Pacific there can be revolutions in song. These patterns of change mean that humpback song has to be culture. Humpback whale cultures are broader though, incorporating migration routes and feeding innovations, and they are not the only great whales to have forms of culture.

HUMPBACK WHALE SONG

As noted in Chapter 4, the songs of humpback whales have captivated the general public, musical composers, and scientific experts since their complexity and hierarchical structure were first described in 1971. Male humpbacks produce songs with repeated complex themes. The song is constructed of individual vocal utterances, grouped into phrases, and a sequence of one or more similar phrases is called a theme. Songs consist of four to ten themes arranged in a relatively fixed order. Songs are produced in cyclical bouts that can last for hours. As only males sing, most researchers agree that it has something to do with mating, but whether the songs function to attract females, to allow females to comparison shop between the singing males, or to intimidate other male rivals is still debated by scientists.

SONG CULTURE: EVOLUTION & REVOLUTION

At any one time, all males in a population sing the same song, but the song they sing changes gradually over time. For this to happen, the whales must be listening to each other, and changing their own songs in response, in a process of cultural evolution. In some places the song can change into an entirely new form within less than a year—a revolution, rather than an evolution. Professor Mike Noad and his colleagues at the University of Queensland first documented a song revolution off eastern Australia. In 1996 a single whale was heard singing a dramatically different new song, and by the end of 1997 virtually all the whales off eastern Australia were singing it. This "new" song was actually the one being sung by the western Australian population at the time. It seems that a few whales from this western population ended up, whether by accident or design, swimming up the eastern seaboard from their shared feeding grounds in Antarctica, and their song proved irresistible to the east-coasters. It has since been discovered that these revolution events propagate across the South Pacific, from the east coast of Australia to French Polynesia, some 3,700 miles (6,000 km) spanning several humpback breeding grounds. A song that was heard off eastern Australia and New Caledonia in 2002 made it to the Cook Islands in 2003, and French Polynesia in 2004. By then, it had been superseded by a different song on the east coast of Australia. So songs continually propagate from east to west in waves. We have no idea why, and understanding the cultural evolution of song over these huge areas is a research challenge for the future.

HUMPBACKS ARE NOT THE ONLY SINGERS

Bowhead, fin, and blue whales are other mysticetes that sing, albeit in more simple form. The blue whale song, like that of the fin whale, contains a single theme, which is repeated. Each phrase of the theme is made up of one to five different units in a prescribed sequence. The songs are too low in pitch for us to hear properly unaided. These whales can sing for days, taking only short breaks to breathe. Currently, blue whale scientists recognize 11 song types worldwide from different ocean basins, suggesting distinct populations. All whales within a population sing virtually the same song. Unlike in humpbacks, the structures of these songs are remarkably stable over decades, except for the extraordinary discovery that blue whale songs in every single one of the world's oceans have gradually, but steadily, got lower in pitch over the last 30 years. Whale scientists have produced a range of far-fetched explanations for this, but little evidence. Nonetheless, this high conformity and change in unison over weeks and years means that cultural evolution is taking place in blue whale song. Bowheads sing repeated phrases in cycles that can last up to 30 minutes, and in any one place the whales will be singing multiple songs, in contrast to both the blues and the humpbacks. The songs change completely between years, however, so again they show cultural evolution.

MULTIPLE TRADITIONS

Song is not the only cultural inheritance of mysticetes. Many mysticetes learn their migration routes from their mothers and this has important consequences for their population structure and conservation. Born in warm-water breeding grounds, young whales follow their mothers away from the equator to summer feeding

Changes in blue whale song pitch

The song of the blue whale in the northeast Pacific has steadily deepened in tonal frequency. After McDonald et al. (2009).

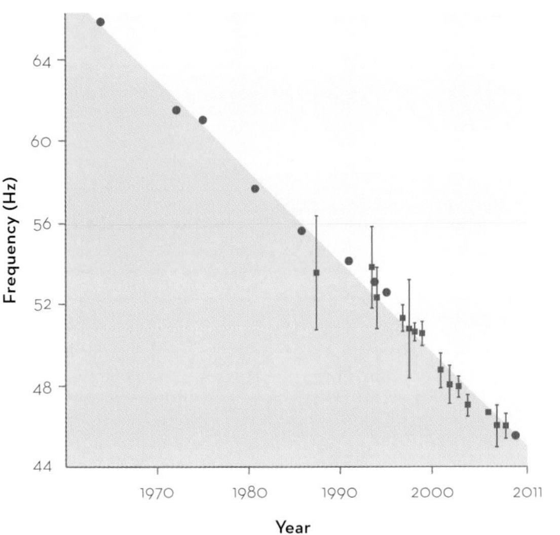

Humpback song transmission in the South Pacific

Biologists in Australia recorded songs in six populations of humpbacks. In any given year, all the males sing the same song but the songs change from year to year, suggesting cultural transmission. The distinct song types, represented by the different colors, moved from west to east in the period of study. Two colors within the same year and location indicate that both song types were present; the banding indicates the season (early, mid-, or late) when the new song type was recorded. Pale gray indicates that no data are available. After Garland et al. (2011).

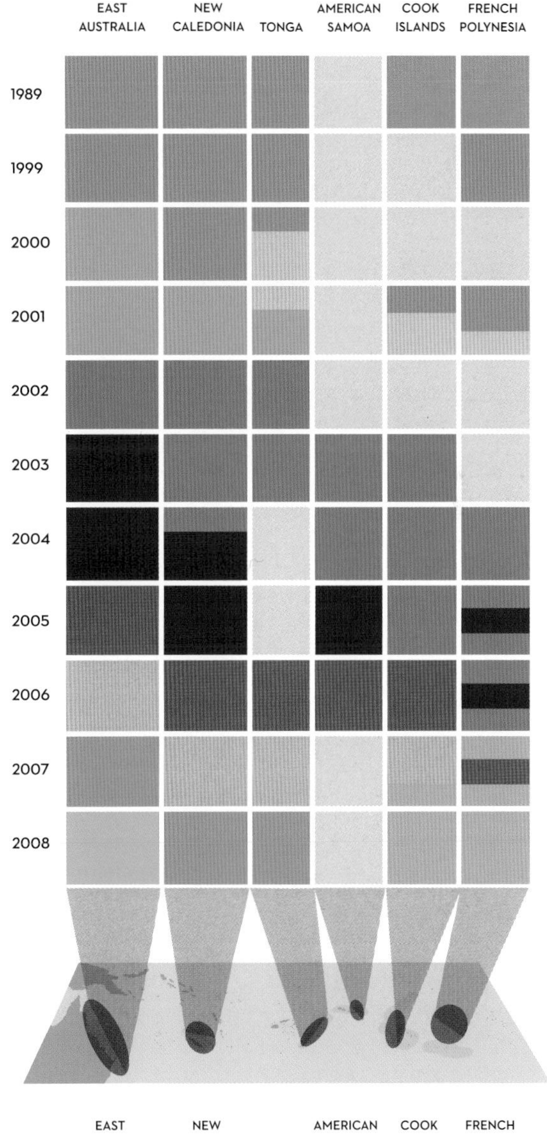

areas in their first year of life. They learn the route from their mother and repeat it pretty much every spring and fall for the rest of their lives. At a population level, this results in population divisions that can be detected in the genetic profile of the species. This conservatism makes sense. Suppose you, a humpback whale, were born on Silver Bank in the West Indies and after a few months you follow your mother to Newfoundland, learning to feed there with her: the good places, the different techniques for lunging on krill, capelin, or squid. When you strike out on your own, would you risk swimming off in an unknown direction? In the deep ocean there are many more chances to fail than there are to succeed, so why take the risk? Finally, it seems that some species are also able to respond to ecological change with foraging innovations that spread through a given community. We will illustrate this in the case study on pages 132–3, which documents the spread of a feeding innovation through the humpback whales of New England.

Top Blue whales, Santa Barbara Channel, California. Populations of blue whales each appear to have their own characteristic song, and researchers have documented how these songs are changing pitch over decades.

Above We now have evidence that humpback whales learn their migration routes from their mothers, while male humpbacks learn their song from other males in the same population, and both males and females can learn new feeding tricks from each other ... but there remains much more to discover.

Annual migration routes of mysticetes

Migrating between warm, low-latitude waters (winter breeding grounds), and cooler, high-latitude waters (summer feeding grounds), northern and southern populations probably never meet.

BLUE WHALE MIGRATION ROUTES

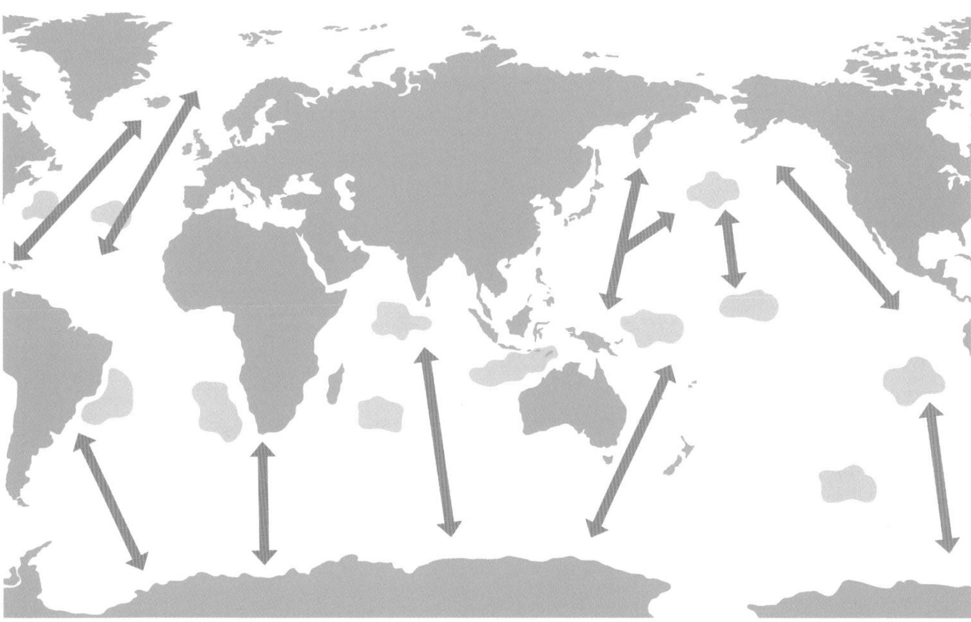

HUMPBACK WHALE MIGRATION ROUTES

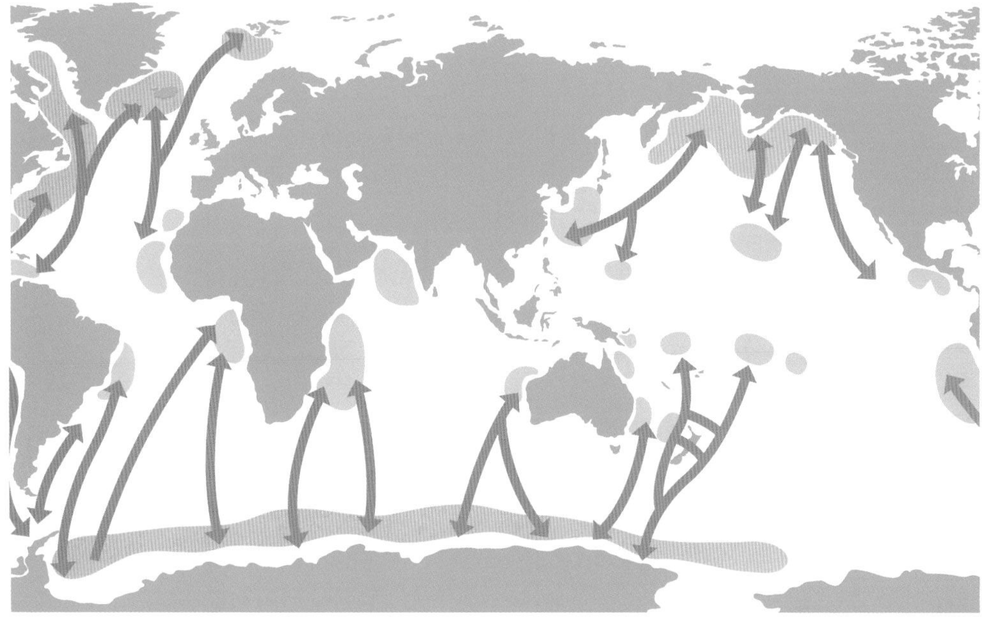

Winter breeding areas · Summer feeding areas · Migration routes

CASE STUDY: **KICK FEEDING IN HUMPBACK WHALES**

Studying the role of cultural transmission in great whale societies is a major challenge—we will likely never be able to run experiments on their learning capacities. All is not lost though. Now that long-term studies of many populations are maturing, and computational analysis techniques are increasing in power, we are able to use simple data collected over decades to track the spread of behavior.

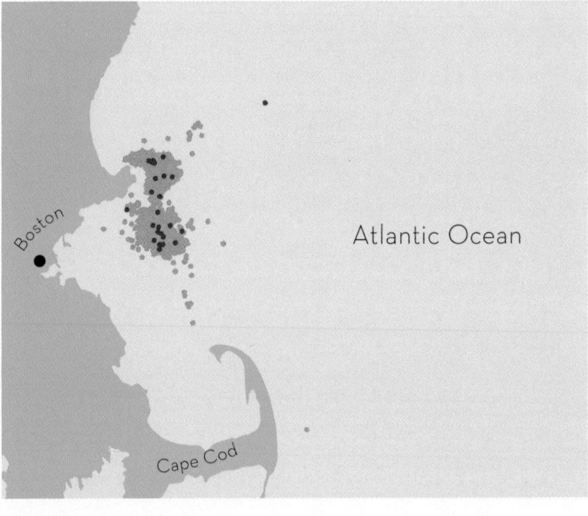

DATA COLLECTION 1981–3

● Kick feeding ● Whales observed feeding without kick

1 KICK FEEDING APPEARS

In the waters off Cape Cod, Massachusetts in the northeast of the United States, researchers led by Mason Weinrich have collected observations of humpback whales from whale-watching vessels every summer for decades. The whales show up to take advantage of local prey abundance, so they feed a lot. In 1980, the researchers noticed a new technique, called kick feeding, used by just one of the 150 feeding whales they observed that year. In kick feeding, animals flick their tails as they submerge, creating a "kick" that disturbs the water and creates clouds of bubbles just below the surface.

2 DATA ARE COLLECTED OVER DECADES

In 1981, two animals were observed using the kick feeding technique. By 1989, 42 of the 83 whales seen surface feeding did it. Kick feeding had become common through the population in just nine years. How had it spread? Weinrich continued to collect data over 30 years on which animals showed this behavior, as well as where and when it was seen. Since originally reported in 1980, this behavior has persisted in the population to the present day, but it has waxed and waned in popularity, judging from the rates at which it has been observed over the intervening years.

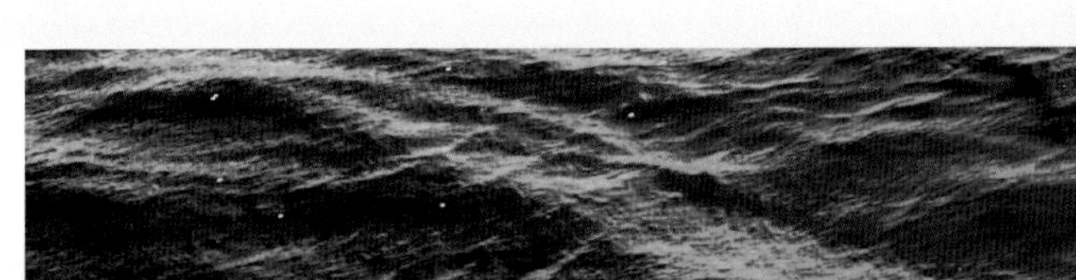

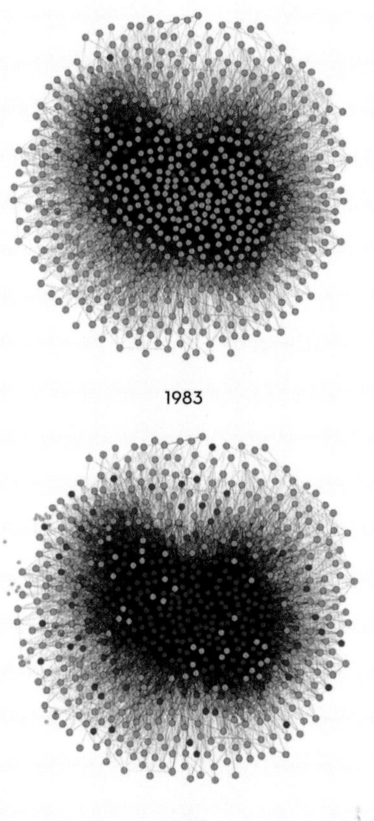

These diagrams represent the social network of the population laid out so that whales with lots of connections are drawn to the center. Kick feeding spread through well-connected individuals near the network center (whales seen kick feeding are shown in red).

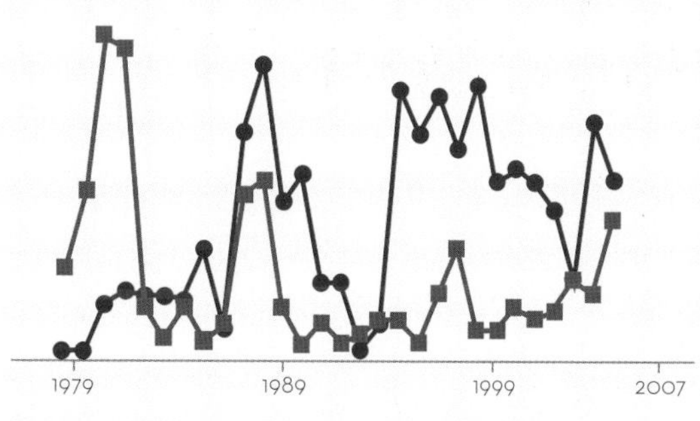

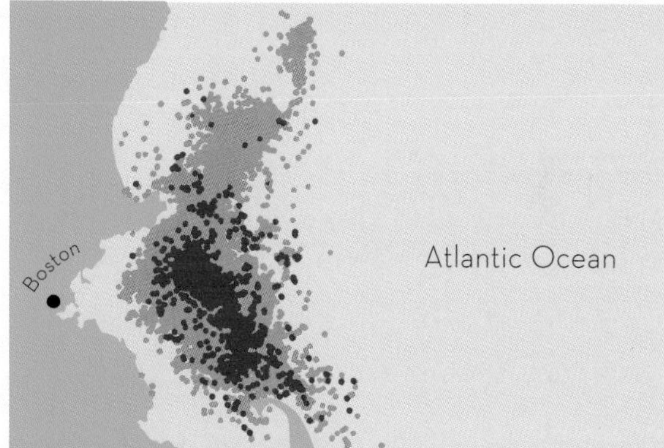

DATA COLLECTION 1981–2007 (CUMULATIVE)

Kick feeding Whales observed feeding without kick

3 APPLICATION OF NEW MODELING TECHNIQUES

Weinrich teamed up with a student, Jenny Allen, and two biologists based at the Sea Mammal Research Unit in the University of St. Andrews, UK, Will Hoppitt and Luke Rendell. Together they used a statistical technique called "network-based diffusion analysis" to analyze the spread of kick feeding. This works on the simple notion that you should be more likely to learn something from another individual the more time you spend with that individual, if culture is indeed involved. If there is no social influence, then it shouldn't matter how often you see others do it. When the analysis was complete, the evidence ratios showed it was literally tens of thousands of times more likely that culture was involved than it was not.

4 UNDERSTANDING THE CULTURAL SPREAD OF A NEW FEEDING TECHNIQUE

So what is the deal with kick feeding? The analysis also showed that the incidence of kick feeding in the population waxed and waned roughly in line with how much of a particular prey species, a small fish called sand lance, was around at the time. Kick feeding appears to be a specialization for capturing sand lance. The appearance of kick feeding coincided with a major shift in prey availability. Herring stocks, the whales' previous favorite, collapsed in the late 1970s and sand lance populations boomed in the early 1980s. Thus the study showed how humpback populations could use culture to adapt flexibly to a changing environment.

MOTHER CULTURES: SPERM WHALES & KILLER WHALES

Compared to the dolphins and mysticetes the larger odontocetes have a very different social structure, based on motherhood. Matrilineal groups of sperm whales and killer whales have their own vocal dialects, which are clearly cultural in nature, and many other group-specific behaviors that are likely culturally transmitted. We also see the spread of new feeding methods, especially in a cultural radiation of killer whale diet specializations. Cultural conservatism in the feeding habits of killer whales is so strong that it has dramatic impacts on the genetic evolution of species.

MOTHER CULTURES

Communities organized along maternal lines provide a strong foundation for culture. The societies of some of the larger odontocetes species are strongly structured, with a focus on mothers—most females spend their lives in the same social group as their mother while both are alive. The fish-eating killer whales off the northwest coast of North America are a rare example where both males and females spend their lives in the same groups they are born into. For sperm whales the basis of society is again the matrilineal unit, but males leave at puberty. Females within units suckle and babysit each other's young and defend themselves communally.

CLICKS & CULTURE IN SPERM WHALES

Sperm whales regularly produce short patterns of clicks, repeated in stereotyped patterns. Called "codas" (because they were first described as being made at the end of a long dive), these patterns can be given shorthand descriptive names that give some idea of their form. For example, "5R" is used for five regularly spaced clicks, or "3+1" for three regular clicks with a longer pause before the final one In some regions, social units of sperm whales can be grouped together into "vocal clans" according to the kinds of codas they make. Five vocal clans have been documented in the South Pacific. Genetics does not drive these dialects—vocal clans are cultural entities.

Cultural diversity in sperm whale codas

The rhythmic pattern of clicks is indicated by the spacing of the dots, so a 3+1, characteristic of Mediterranean dialects, is "click-click-click-pause-click." Across the globe, sperm whales in different places make different types of codas—off Dominica, for example, most groups share a distinctive 1+1+3 pattern (below left). However, in some places, like the Galapagos, groups with different dialects co-exist—the societies are multicultural (below right).

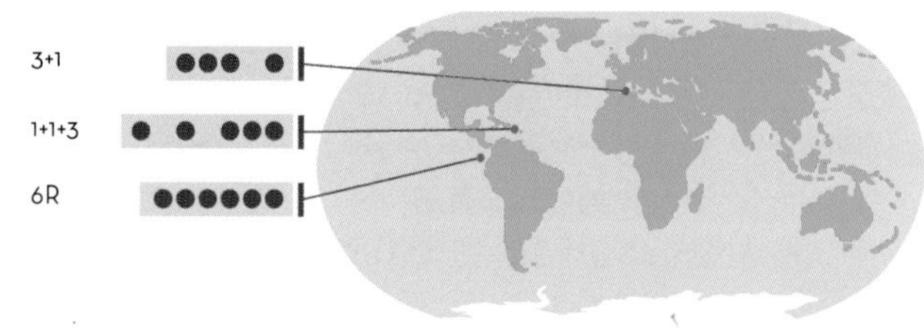

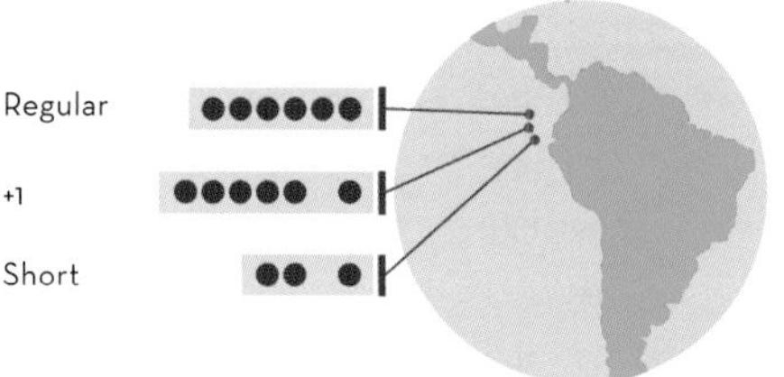

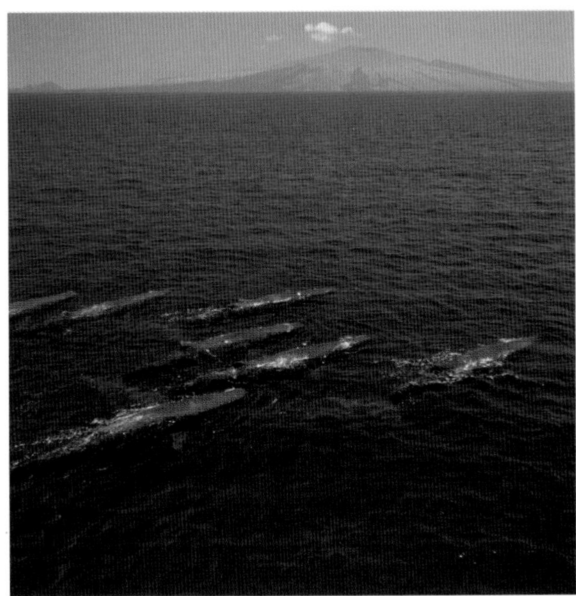

Left A group of female sperm whales and their young rest at the surface off Pinta Island in the northern reaches of the Galapagos archipelago. During their long juvenile periods it seems that young sperm whales learn their vocal dialects and their foraging tactics from their mother and her relatives—hence "mother cultures."

Around the Galapagos Islands studies have shown that different clans have characteristic ways of behaving. One stays close to the islands, another keeps about 6 miles (10 km) farther from land. Groups from one clan moved in a more zigzag fashion while the groups from another clan moved in straighter lines. As a result, groups from the different clans had different feeding success—so the cultural inheritance a young sperm whale receives from its mother's group has direct effects on its life chances.

CULTURAL DIVERSITY & CONSERVATISM IN KILLER WHALES

As in sperm whales, social groups of killer whales have their own vocal dialects. These are learned—in captivity young killer whales will adopt the calls of the animals they are housed with, not the population they came from. In the wild they learn their calls from their mothers and other members of their matrilineal group. Scientists have shown how calls change subtly over time, but also how groups keep up with the changes in synchrony. Vocal dialect is a key part of a killer whale's cultural inheritance, identifying them within their social context, but it is just one aspect. Best studied are the killer whales of the northwest Pacific, where there are three distinct "ecotypes"—forms with distinct ways of life, defined primarily by their food. The inshore fish-eaters eat salmon, but there are mammal-eaters and offshore fish-eaters that prey on deepwater fish, especially sharks, and all using the same waters. In the Antarctic there are at least four ecotypes, specializing on minke whales, seals, penguins, and fish respectively. While the killer whales of the world eat a very wide variety of prey, from blue whales to sting rays to herring, each individual killer whale and each ecotype of killer whales has a much more restricted diet, and sticks to it with extraordinary conservatism. In 1970 when three mammal-eating whales were captured alive off British Columbia for the display industry and housed with fish-eating whales, no one knew that there were different ecotypes of killer whale. For 75 days they were provided with salmon, but the mammal-eaters refused them. One died. Four days later the other two began to eat the fish, but they reverted to mammal food on being returned to the wild after a few months.

INTERACTIONS WITH HUMAN ACTIVITIES

As with the dolphins, the cultural learning capabilities of killer and sperm whales have led to interactions with humans. Off southeast Alaska, for example, male sperm whales have learned how to take fish from longlines—with remarkable dexterity, they use their teeth to pluck and twang the tense line until the dead hooked fish are dislodged, making an easy meal. This behavior has spread from a single point of origin in a way that fits the "wave-of-advance" model used by social scientists to map cultural change in human societies.

Perhaps the most remarkable interaction is the historical cooperation between killer whales and shore-based whalers off Twofold Bay in southeast Australia between about 1835 and 1930 (see also Chapter 1). When a humpback came along, some of the killer whales would swim over to the whaling station and start jumping out of the water (known as " breaching") to alert the whalers. Others would herd the humpback into shallow waters, into the path of the whaling vessel which could relatively easily harpoon the harassed humpback. The whalers then moored the dead whale to the seabed, so the killer whales could pry open its mouth to get to their preferred delicacy, the tongue. After 24 hours, the human whalers would return and retrieve the carcass for its oil.

WHY IS CETACEAN CULTURE IMPORTANT?

Understanding cetacean culture is important. It helps us understand how cetaceans come to behave the way they do, and how they have evolved. It has implications for conservation, because their culture can affect how they respond to human impacts—from fishing cooperatives, to longline depredation, recovery from whaling, and habitats being altered by climate change. Understanding how culture evolved in the ocean can potentially shed light on our own origins. Finally, understanding whales and dolphins as cultural beings could potentially impact our moral perspectives on how we treat them—and other nonhuman cultures.

GENE-CULTURE CO-EVOLUTION

Human societies with a history of dairy farming have genes that make us lactose-tolerant as adults. This is called gene-culture co-evolution and has been a major force in human evolution. But perhaps we are not the only species in which it occurs. What is going on with all the different killer whale ecotypes? Are they species, or subspecies, or races? The differences between the ecotypes are profound—genetics show that the lineages have been separate for many, many generations. In this scenario, we see killer whale evolution as a slowly changing patchwork of ecotypes, splitting, evolving, specializing, and disappearing: a dynamic, vivid pattern of culturally driven diversification that also drives the evolution of genetic patterns. This unusual mode of evolution is driven by the importance of culture in the lives of killer whales. Killer whale evolution is bound up with killer whale culture.

CULTURE & CONSERVATION: IT'S COMPLICATED

Culture complicates conservation in many ways. Horizontal cultures—learned from peers—can propel populations into problems with humans, but they can also help them deal with the problems. If a trick for taking fish off a longline had to be invented from scratch by each new animal, there would be many fewer animals doing it, perhaps just the occasional very smart or lucky creature. In a cultural species though, a single innovation can change the behavior of a whole population. We doubt it is coincidence that the species overwhelmingly implicated in taking of fish from nets or lines are the bottlenose dolphin, the killer whale, and the sperm whale—three species where the evidence for culture is strongest. On the other hand, vertical cultures—learned from parents—can be disrupted by human activities and may inhibit adaptation to them. Southern right whales, for example, learn their migration routes from their mothers, and tend to be quite conservative about changing them (one can understand why, given the multitudinous ways there are to get lost in a featureless ocean). Whaling, however, severely depleted their numbers, and it seems that the cultural knowledge of some migration routes has been lost, so that some populations are struggling to recover even while others appear to be doing well. Conservatism in things like migration routes means that we cannot take recolonizations of suitable habitat for granted once the cultural knowledge is lost.

CULTURE & CONSERVATION: POLICY

The questions around cetacean culture can have direct consequences for conservation policy. Take the status of the southern community of fish-eating killer whales off the Pacific coast of North America. In 2002, the US National Marine Fisheries Service ruled that they were not a "distinct population segment" and so did not merit special protection. This prompted law suits, and in 2005 the US government reversed its decision, "based on evaluation of ecological setting, range, genetic differentiation, behavioral and cultural diversity," and the whales were listed as distinct and endangered. The pages of the US Federal Register

Above Southern right whale mother and calf in the De Hoop Marine Protected Area off Western Cape Province, South Africa. Recent studies have recovered genetic signatures that tell us young right whales learn their migration routes from their mothers. Since getting your migration wrong is a serious problem, this learning is highly conservative, which impacts population recovery post-whaling when knowledge of routes has been lost.

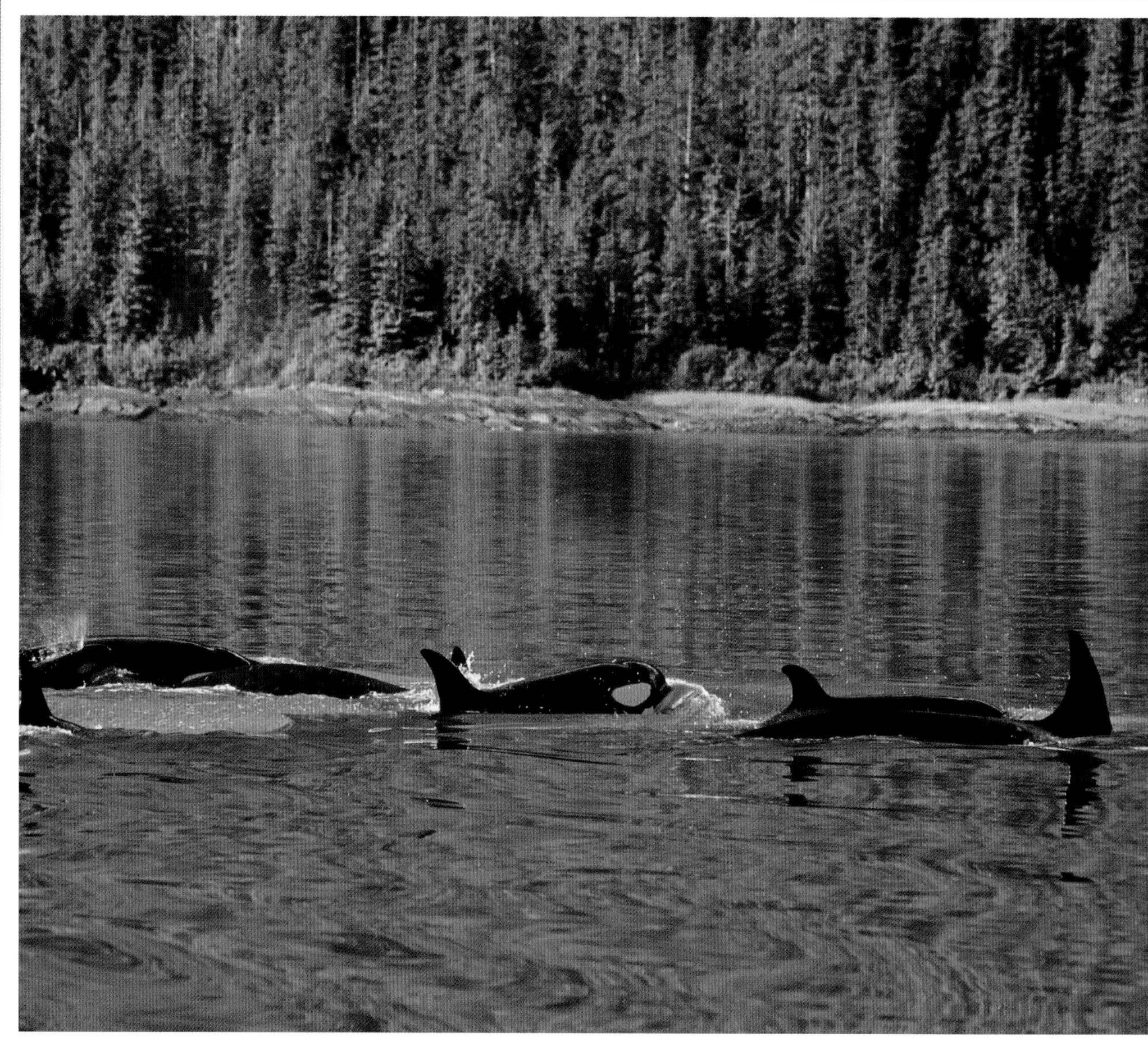

announcing the decision contained the statement "there are differences in cultural traditions, and the Southern Residents may have unique knowledge of the timing and location of salmon runs in the southern part of the range of North Pacific residents." While we can always wish for more knowledge and more data to inform our policy, in our societies we need to be making decisions right now about how we treat, conserve, and manage populations of whales and dolphins. If we don't get the decisions right, with our limited understanding of the forms of culture present in these populations, our future whale scientists won't be biologists, they will be paleontologists.

Above An inshore, fish-eating population of killer whales, Alaska. Each killer whale population appears to carry specific knowledge about food and habitat that might have taken many generations to accumulate.

WHALE CULTURE, HUMAN CULTURE

Comparing how culture has evolved in different ways in different species can tell us about how it might have evolved in human prehistory—it can help us fill in the early chapters of our own story. Cetaceans have a variety of social structures, inhabit a broad range of ecological niches, and have varying cultural systems. Thus studying this diversity offers a greatly increased understanding of how social and ecological factors

impact the evolution of culture, and the capacity for culture. Such understanding will inform those seeking to comprehend humanity's own biological and cultural evolutionary history.

Above Killer whales feed opportunistically alongside fishing activity in a Norwegian fjord. The way cetaceans can adapt by learning from each other how to exploit new feeding opportunities created by humans can lead to rapid expansions of conflict between whales and fisheries.

OUR MULTICULTURAL WORLD

Understanding our world as one in which different forms of culture exist might also lead to reflections on our responsibilities. We are living in a time when the relationship between human economic culture and the ecosystem on which it depends is in question as never before, and alternative perspectives about the relationship between humans and the natural world are being increasingly valued. What rights do other species have? Should we understand members of some cetacean species as persons, with the full moral implications that status brings? Culture is not included directly in the usual requirements for personhood, but it can be seen as the context in which personhood is expressed. Studying culture in cetaceans could therefore directly inform some highly contentious moral debates.

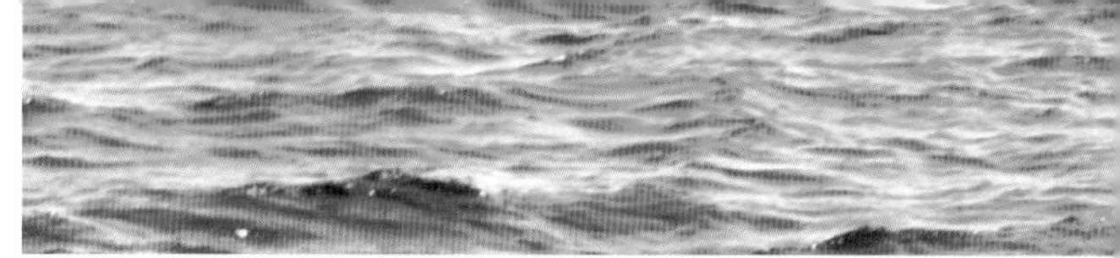

GONE FISHING

Janet Mann

People often ask: "Why can't dolphins raised in captivity survive in the wild?" There are many reasons why attempts to release captive-reared dolphins into the wild have failed. One key reason is that they have great difficulty catching fish. Keiko, the killer whale made famous from the blockbuster film *Free Willy* continued to beg for fish from humans and died despite our best efforts following his release in Icelandic-Norwegian waters. Even with their impressive echolocation and swimming abilities, which would undoubtedly have been affected by prolonged periods in confinement (that is, sensory deprivation), dolphins and whales often feed on diverse types of prey or must learn specific hunting techniques according to family tradition (for example, killer whales) or location. Although sperm whales, pilot whales, and many species of beaked whales specialize on squid, smaller dolphin species might feed on dozens of prey types. While it seems basic enough to swim after a fish and catch it in your jaws, it is harder than it seems. Each type of fish has distinct ways of avoiding being caught, called anti-predation strategy, and dolphins must overcome these. Shark Bay bottlenose dolphins provide a window into this problem since their foraging behavior has been studied for more than 30 years. Shark Bay dolphins exhibit more than 20 distinct hunting tactics and most dolphins specialize in a small subset of these.

Above Snacking involves chasing small fish to the surface and swimming belly-up near the surface to trap them. Although dolphins of all ages snack, this foraging method is particularly popular among young calves and it is the first foraging technique Shark Bay calves are observed engaging in. Here more than one young dolphin is snacking, what we call a "snack party." Sometimes they risk running into each other while chasing fish.

HUNTING & FORAGING TECHNIQUES

If we begin at the beginning, a young calf, weeks old, is entirely dependent on his or her mother for food, and will continue to nurse for three to eight years more (although typically until three or four years old). But calves attempt to catch small fish in the first months of life, and although they are frequently unsuccessful, by 3–4 months, they are catching small fish. Early on, calves engage in a hunting behavior called snacking, in which the dolphin chases a fish (even quite tiny fish, less than an inch long), to the surface and the fish is then trapped there while the dolphin swims belly-up, chasing it. Why belly-up? Well, a dolphin's vision is best beneath their heads (looking down), so by being belly-up, they can see the fish quite well—especially since the fish is now back-lit. For young calves, they might be able to hone their echolocation skills in concert with the visual image of the fish. Sometimes the fish jumps in air to escape, which is when the dolphins snaps and catches it. Garfish are the most common prey species that are "snacked" on and adults occasionally snack too, sometimes with garfish reaching more than a foot and a half in length.

Calves also closely observe what older individuals, especially their mothers, are hunting and catching. One behavior, called "fish inspections," occurs when one dolphin (usually younger) follows and "inspects" the fish caught by another individual. It is not uncommon to see several calves and juveniles gather around an adult

Above When dolphins catch their prey, particularly a big fish like this pink snapper, others come to inspect the catch. Young dolphins are more likely to inspect the fish catches of older dolphins than the reverse, suggesting that young are learning which fish species are good to eat.

who has caught an impressively large fish. This looks like begging, but it isn't. The "inspector" receives nothing.

By the second year of life, calves have been trying out the foraging tactics their mothers use, albeit with some interesting exceptions. Sponge tool-use takes years to acquire, with most calves getting their own sponge at age two or three and becoming proficient at sponge-foraging much later (see Chapter 7). Another challenging hunting behavior is strand-foraging, or beaching, where a dolphin accelerates at high speeds and literally chases its prey, mostly mullet, onto the beach. Sometimes the fish is completely beached and out of reach and the dolphin squiggles or makes a U-turn back into the water. Calves have been observed beaching part-way as early as age three, but not very often until age five or later. This risky technique can obviously leave one stranded, so they tend engage in

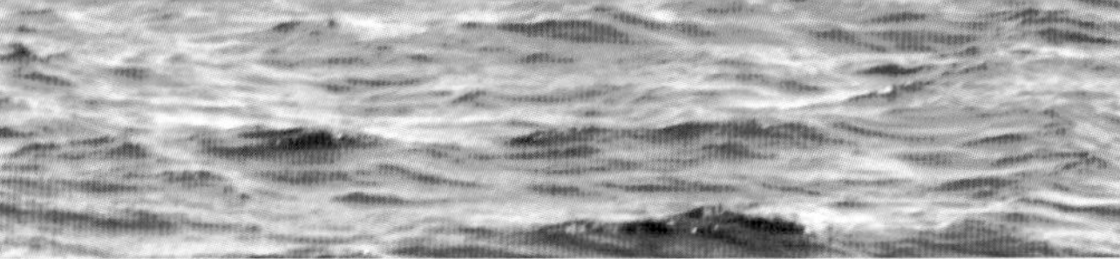

Above Bottlenose dolphin strand-foraging or beaching at Point Peron in Shark Bay, Australia. Dolphins beach to trap mullet at the shoreline, a technique offspring learn from their mothers. Others observe the method closely but do not adopt it.

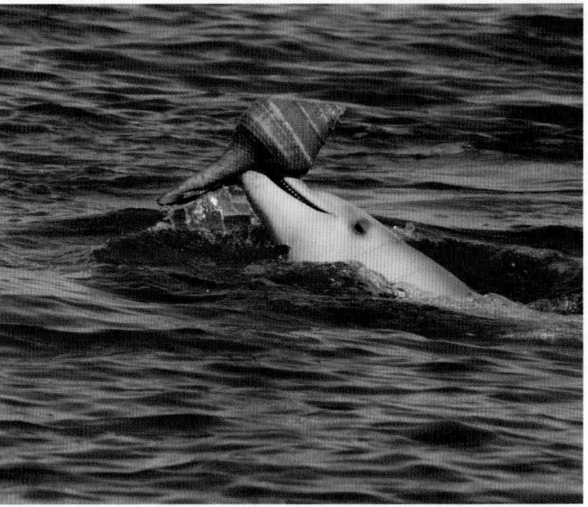

Above Shelling is a behavior seen in Shark Bay in which dolphins retrieve a large trumpeter or bailer shell and lift it out of the water, apparently draining the water and enabling them to get to the fish hiding within. Here, a young female is carrying a trumpeter shell, home to the largest snail in the world. We don't think she is after escargot, but possibly a fish.

this behavior with an incoming tide—when the mullet congregate near shore. About seven dolphins have been seen beaching in Shark Bay, all born to one matriline.

A rare tactic that calves do not use is shelling, when a dolphin either chases a fish, or finds a fish in a large trumpeter or bailer shell in which the mollusk is no longer living. Because the fish is unwilling to vacate the shell voluntarily, the dolphin makes it easy on him by raising the shell in its beak clear of the water, thus draining the shell and fish into its mouth. This dramatic behavior entails raising a very heavy shell clear of the water. We still know little about the behavior—for example, how the fish ends up in the shell and whether dolphins re-visit the same shells. Only a handful of dolphins have been observed shelling, although this behavior also seems to pass from mother to offspring.

Other hunting techniques are more common, but also require skill and practice. Bottom-grubbing is observed in shallow beds of sea grass where the dolphins chase and poke around in the sea grass to find fish. Although they echolocate, sea grass scatters the sonar, so the primary method is to scare up the fish so they can see it and catch it. Since sea grass beds are great places for fish to hide, there are plenty of prey there and I have watched many dolphins using this habitat in Shark Bay, which has some of the largest and richest sea-grass beds in the world.

Cuttlefish are a favored prey item for dolphins, and are easy for humans to see because their ink colors the water when they are disturbed. But cuttlefish have a large central bone which the dolphins don't swallow. Although it is unclear how they do this, they seem to bite down on the cuttlefish in such a way that the bone squirts out cleanly, leaving one or even no set of dolphin toothmarks on the bone. Catching prey is just one step, handling prey that have spines, barbs, large heads, or bones can be a delicate operation.

One of the most dramatic behaviors we observe in Shark Bay is golden trevally hunting. These large delicious fish reach more than three feet (over a meter) in length. They are very fast, and only a few dolphins seem capable of catching them. Wedges, an infamous female dolphin, would catch one of these fish every couple of hours—and they were sometimes more than half her body length.

The chase is dramatic, with extensive leaping in every direction until Wedges catches the trevally. First, she takes it down to the seafloor, presumably to break its neck so it won't escape from her narrow jaws. Then makes a beeline for shallow water, a yard or two deep, where she can snap the head off and break the fish up into pieces she can swallow. This often takes an hour or so, during which time she attracts other dolphins (who watch) and sharks, including large tiger sharks,

Above Wedges was the first dolphin seen to hunt golden trevally, but another family, the Puck family, also catches trevally—usually when Wedges is around, suggesting that they picked it up from her. These fish can be half the dolphin's body length.

that attempt to steal the fish. Her calves don't have an easy time of it either, since they are unable to nurse during that period, and must avoid the sharks too. Wedges is capable, however, of defending her prey.

As we describe in Chapter 7, which discusses tool use, only a subset of dolphins appears to engage in these behaviors and they are often become specialists, favoring one technique over all others. Several factors are likely to favor this pattern. First, the marine environment can be quite diverse in terms of the types of prey that are available and the different habitats. These change by season as well. With so many types of prey, it might be better to become skilled at catching a few, than trying to hunt all kinds. Second, such skill takes learning and practice and who better to learn from than your mother? Most of the techniques are learned from the mother, although some they might pick up on their own or from other dolphins. This also means that dolphins might learn which fish should be avoided because of their stinging spines or toxins—as in puffer fish. Some of the hunting tactics seem easy to learn, such as snacking, while others, like sponging, take years to master. Each tactic brings its own challenges, for example finding the right sponge tool (see Chapter 7), finding the right kind of fish, chasing and catching the fish, handling the prey, and even avoiding sharks in the process.

7 CETACEAN TOOL USE

Eric Patterson & Janet Mann

Untitled, Humpback Whale Mother and Calf
Kingdom of Tonga, 2005

"I often found this mother and calf pair resting near our dive base in water as shallow as six meters. At times the mother would 'walk' by pushing her pectoral fins against the sandy bottom."—Bryant Austin

DEFINING TOOL USE

Cetaceans have long fascinated scientists and the public alike. Dating back to Charles Darwin, people have also been captivated by animal tool use. Although we are most familiar with tool use in primates, the few known examples of cetacean tool use have attracted immense interest, in part because cetaceans have streamlined physiques and their marine environment makes tool use improbable. In this chapter, we discuss how scientists define tool use, explore cases of cetacean tool use, and finally consider how cetacean tool use helps us to understand the factors that promote and impede the evolution of tool use more broadly.

Above Chimpanzee using two rocks as a hammer and an anvil to crack a hard nut, Guinea, West Africa.

When Jane Goodall first reported that chimpanzees use sticks to probe for termites, her mentor, the famous anthropologist Louis Leakey, said "Now we must redefine 'tool,' redefine 'man,' or accept chimpanzees as humans." In this timeless quote, Leakey struck on a controversial issue faced by those that study tool use: its definition. Since then, scientists have hotly debated the definition of tool use. In fact, in their book *Animal Tool Behavior* (2011) cataloging the known cases of animal tool use, Robert Shumaker, Kristina Walkup, and Benjamin Beck dedicated almost the entire first chapter to defining tool use and manufacture. Most definitions of tool use are lengthy and somewhat convoluted. Here, we adopt the definition of Shumaker and his colleagues with one small modification (italicized word): "the *conditional* external employment of an unattached or manipulable attached environmental object to alter more efficiently the form, position, or condition of another object, another organism, or the user itself, when the user holds and directly manipulates the tool during or prior to use and is responsible for the proper and effective orientation of the tool."

We add the word "conditional" to emphasize that, to us, tool use involves object use in a specific context. It is this conditionality that makes tool use of interest, especially in terms of cognition. Some forms of tool use, such as nut-cracking in chimpanzees, involve planning, extensive learning, and understanding object properties. Here, chimps bring only hard-to-crack nuts to specific sites that have anvils (appropriately shaped and sized rocks) and hammers (large stones or logs) available. Young chimps study their mothers and other tool users closely before attempting the technique themselves. It takes years to become proficient and some never get it right. Clearly some forms of tool use require a little extra brain power, but that does not mean that all large-brained animals use tools or that all tool users have large brains. Many dolphin species have relatively large brains, often exceeding those of primates, but tool use is rare in dolphins probably because tools are usually not needed to solve the problems dolphins, and most cetaceans, face. With this in mind, we explore when tools do seem to help cetaceans solve problems.

However, before doing so we must take account of the fact that studying tool use in the ocean is quite different from studying tool use on land. Cetaceans are more often out of sight than in, meaning we only get a short glimpse of their tool-use behavior—that is if it occurs near the surface. Even when observation conditions are perfect, researchers rarely venture beyond coastal, shallow, and surface waters, so we know very little about what is happening far offshore or deep below. Perhaps these difficulties explain why most scientists interested in tool use study birds and primates. Nonetheless, countless hours at sea have uncovered several intriguing examples of cetacean tool use, which we summarize in the table below and further describe in the following pages.

Known cases of cetacean tool use or tool-use-like behavior

There are only eight known cases of tool use among wild cetaceans, with most being exhibited by delphinids, or dolphins, in a foraging context. (The killer whale, or orca, is a member of the Delphinid family.)

SPECIES	DESCRIPTION OF BEHAVIOR	CONTEXT
Amazon river dolphin	Carry and hold objects	Social
Indo-Pacific humpback dolphin	Throw sea shells	Social
Irrawaddy dolphin	Squirt water jets during foraging	Foraging
Humpback whale	Encircle prey with bubble nets	Foraging
Common bottlenose dolphin	Encircle prey with mud plumes	Foraging
Killer whale	Wash prey off ice floes with waves	Foraging
Australian humpback dolphin	Carry marine sponge	Unknown
Indo-Pacific bottlenose dolphin	Wear sponges for protection during foraging	Foraging

SOCIAL TOOL USE

Among cetaceans, there are only two known cases of tool use, or tool-use-like behavior, that occur in a social context. The first involves object waving and may be a form of mate attraction. The second involves object throwing, but here the function is less clear, making it difficult for us to decide whether this is tool use or not.

OBJECT WAVING TO ATTRACT FEMALES

Beyond all else, animals must reproduce. For males, one way to increase reproduction is to attract females, and in one species of cetacean, it seems that males use tools to do so. In the Mamirauá Reserve in Brazil, Amazon river dolphins pick up sticks, stones, mud, and other objects, and wave them in the air. Originally, scientists suspected this was a form of play, but recent evidence shows that most object carriers are adult males in the company of adult females, suggesting that courtship display is the most likely explanation. Whether or not females are "impressed" by such displays and preferentially mate with object carriers is not known, but if they do, this would clearly be beneficial for males and may qualify the behavior as true tool use.

SEA-SHELL THROWING

The second example of possible social tool use by cetaceans is not well understood. In the 1970s, Indo-Pacific humpback dolphins off South Africa's Indian Ocean coast were observed throwing around sea shells in what appeared to be a form of play. Whether or not simply playing with objects constitutes tool use is difficult to say. Since many dolphins toss their fish after catching them, sea-shell tossing might have been some sort of foraging practice, but in playful context.

Below and opposite Amazon river dolphins throwing plant debris, sticks, and stones. Recent evidence identifying most object throwers as males suggests that this may be a form or courtship display in which males use tools to impress females.

FORAGING TOOL USE

Most cases of animal tool use occur during foraging, when the tool is used to extract food or prey from its encasing or refuge. For example, sea otters use rocks to break open shells and crows use twigs to extract larvae from trees. Like these other species, cetaceans also use tools primarily during foraging.

WATER JETS

One possible example of cetacean foraging tool use is the use of water jets by Irrawaddy dolphins. In Southeast Asia, these unusual-looking cetaceans squirt water jets both above and below the surface, apparently to scare and corral fish. However, it is still not quite clear how the jets aid in foraging, if at all. Irrawaddy dolphins—one of the few dolphin species that can survive in freshwater—inhabit fairly murky waters in estuaries and rivers around the Indo-Pacific rim so direct observations of the behavior are few and far between.

BUBBLE CLOUDS & NETS

A better known example of cetacean foraging tool use, and perhaps the only known example among mysticetes, comes from humpback whales. From Alaska and the Gulf of Maine, to the coastal waters of Oman in the Middle East, humpback whales have been observed using bubbles to confuse, encircle, condense, and trap prey. There are two basic forms: bubble clouds and bubble nets. During bubble-cloud feeding, one or a few whales release a large burst of bubbles underwater, and then slowly follow behind in pursuit of the small fish that have startled to the

Setting the trap

Humpback whales sometimes coordinate bubble-net feeding by circling and blowing bubbles below schools of fish. As their bubble net rises, so do the whales until the fish are trapped at the surface and the whales lunge through for an easy meal.

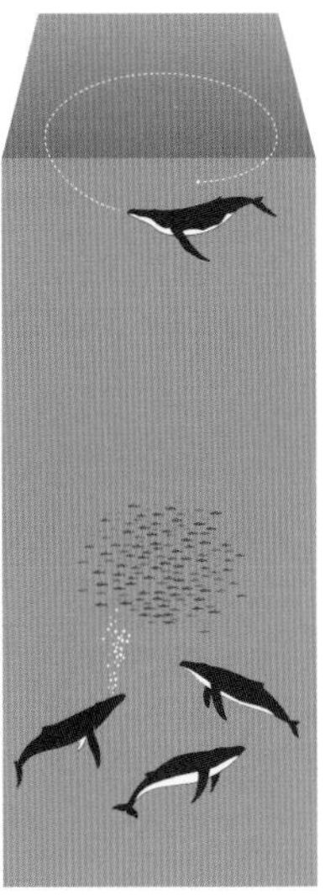
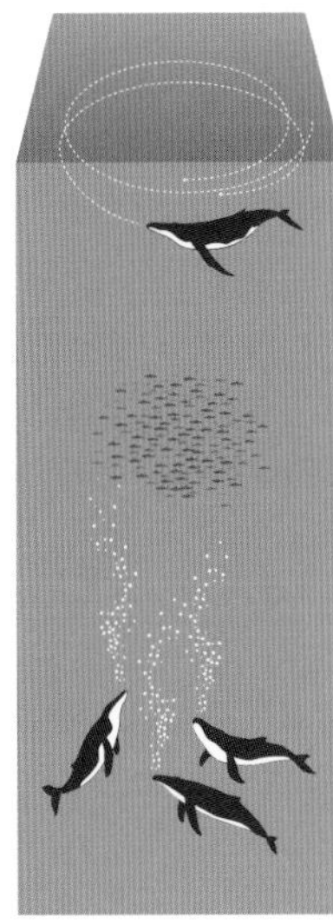

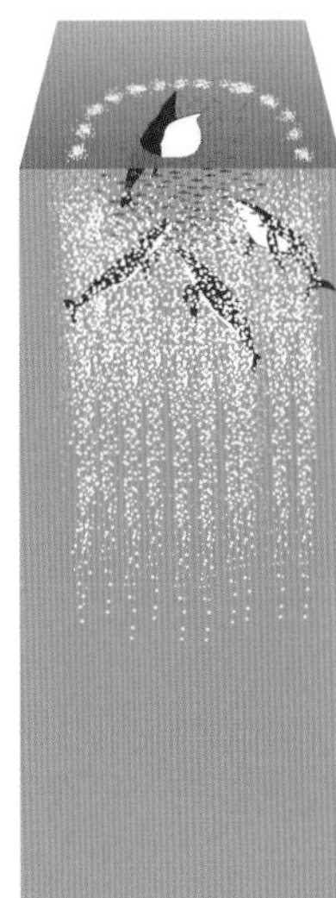

surface. Occasionally more than one cloud is used. In the Gulf of Maine, individual and/or groups of whales sometimes blow several bubble bursts into patterns and feed within the cloudy formation.

Bubble-net foraging takes it to the next level. Here individuals and/or groups of whales blow a continuous stream of bubbles deep below the surface in a circle or semicircle arch around a school of small fish. As the bubbles rise, they concentrate and startle the prey to the surface until the whales lunge through the center and consume large mouthfuls. When more than one whale participates, the animals seem to coordinate their behavior such that all receive an easy meal. Humpback whales fitted with suction-cup tags have revealed two distinct bubble-net methods. Either whales swim in a slow upward spiral to create the bubbly ruse or they perform what is called a double loop, which consists of a corral loop to concentrate the prey, and a capture loop to engulf the prey.

Opposite Irrawaddy river dolphins spitting water. It is thought that these odd-looking dolphins use water jets to help them hunt for fish.

Right Aerial view of humpback whales bubble-net feeding. Humpbacks are the only known species of baleen whales to use tools. Compared to other mysticetes, humpbacks are more agile and have greater maneuverability which may explain why this feeding technique is so effective for them.

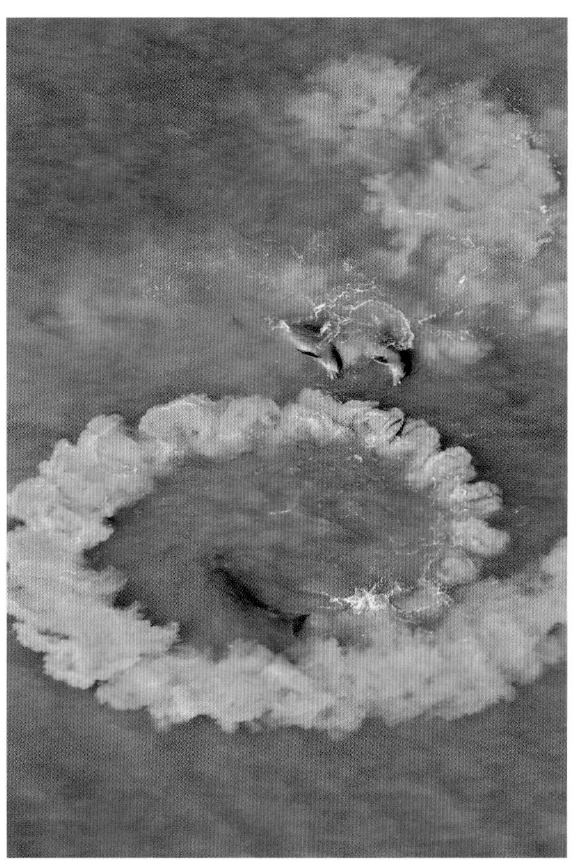

Left Bottlenose dolphins mud-ring foraging. Here dolphins make use of the muddy substrate as a tool to help them condense and capture their prey.

Below Pack ice (or Antarctic type B) killer whales coordinate their tail movements in order to create a wave to wash a Weddell seal off an ice floe.

MUD PLUMES & RINGS

Another encircling method is employed off the coast of Florida by bottlenose dolphins. Here dolphins create mud plumes to encircle prey, usually mullet. As with bubble feeding in humpbacks, there are two variants. In the first, mud-plume foraging, an individual dolphin beats its tail along the muddy bottom in an arch shape, after which the dolphin repositions itself in front of the rising mud plume and rapidly lunges forward to capture the encircled fish. In the second method, mud-ring foraging, dolphins collaborate, although one dolphin usually does all the work. Here one dolphin creates a mud ring by beating its tail on the muddy substrate in a circle around a school of fish. Once the ring is complete, the dolphin and its fellow foragers wait around the perimeter as the mud ring expands inward, eventually causing the fish to leap into the mouths of their captors. How the dolphins decide who does the "dirty" work and/or whether they take turns is not known.

WAVE-WASHING

In cold Antarctic waters, killer whales manipulate their environment in a way that may constitute tool use, a behavior known as wave-washing. The primary prey of these pack ice killer whales are seals but not just any seal, specifically Weddell seals. They search ice floe for prey, and when a Weddell seal is spotted, the whales huddle around the floe and peek above the water to take a glimpse. After sizing up their meal, they swim off to a distance and then surge back toward the floe, side by side, beating their tails in perfect synchrony until nearly headbutting the ice, at which point they roll and dive. The wave that follows crashes onto the floe and, with any luck, washes the seal into the water where the whales wait below. Why these whales so strongly prefer Weddell seals such that they go to great lengths to hunt them remains unknown, but it may indicate that ample prey are available, allowing the whales to be picky eaters.

Below Sponge foraging is the best-known case of cetacean tool use. The bottlenose dolphin in Shark Bay, Western Australia shown here uses an odd, branching sponge to help it forage.

Right More typically, bottlenose dolphins use basket-shaped sponges, as can be seen here.

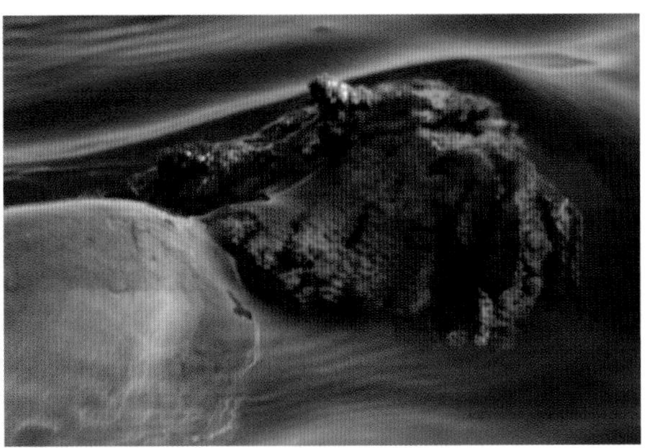

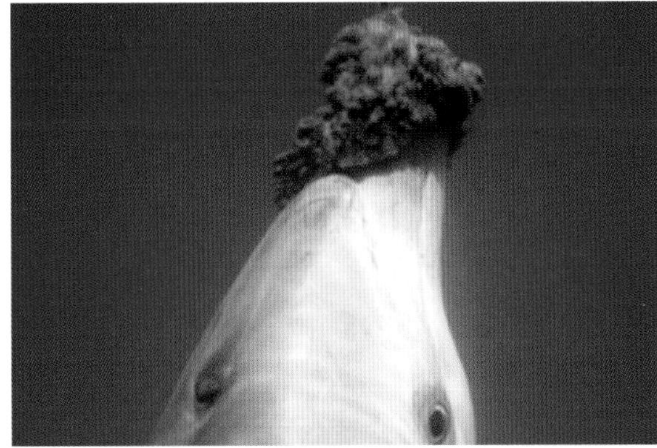

SPONGING

While these previous behaviors are fascinating and may constitute tool use, there is one behavior known in much greater detail that almost no one doubts is true tool use. It is performed by a very small subset of Indo-Pacific bottlenose dolphins in Shark Bay, Western Australia. In 1984, the Shark Bay Dolphin Research Project began at what was then a small fishing camp. During that inaugural year, a fisherman told researcher Rachel Smolker about an odd-looking dolphin with a huge growth on its nose (beak, really), and half a tail fluke. She scoffed until days later she saw a dolphin with a large yellow-orange growth (sponge, really) over its face and, when it dove, half a fluke! She named the dolphin "Half-fluke"—the first known wild tool-using dolphin. In 1989, Half-fluke's two-year old daughter, Demi, picked up a small sponge and at that point, we began to realize that sponging was transmitted from mother to offspring. Through years of research, more than a hundred dolphins have been observed wearing sponges in three locations in the bay, and another curious fact emerged: it is mostly a female behavior, with only a small handful of males engaging in the behavior. Moreover, the females that do it seem to "sponge" relentlessly throughout the day, hunting more than other females and more than male spongers too.

Fully understanding what dolphins are doing with sponges proved difficult, given that sponging occurs in deep channels, beyond what we can usually see. However, on a number of clear and calm days we have observed sponging from the surface: the dolphin cruises along the bottom seafloor, gently disturbing the substrate with her sponge. Then, something darts out, she drops the sponge, chases it, grabs it, and swallows it. After this, she typically retrieves the sponge and starts again. Sometimes the sponger surfaces without her sponge while on the chase, but often somehow finds the same sponge. Dolphins use each sponge for minutes to several hours, before tearing up a new one from the seafloor. Through the help of sponge expert Dr. Jane Fromont (Western Australian Museum) we have determined that spongers use at least five different sponge species.

More recently, researchers in several places around Australia have seen Australian humpback dolphins carrying sponges, but no one quite knows yet why. Sponge foraging has been suggested, but mate attraction is a possibility too.

CASE STUDY: **WHY THE TOOL?**

So what are the spongers after? Why do they need a sponge? Dolphins are exquisite hunters and have extraordinary echolocation abilities. In an underwater experiment, we set out to answer these questions.

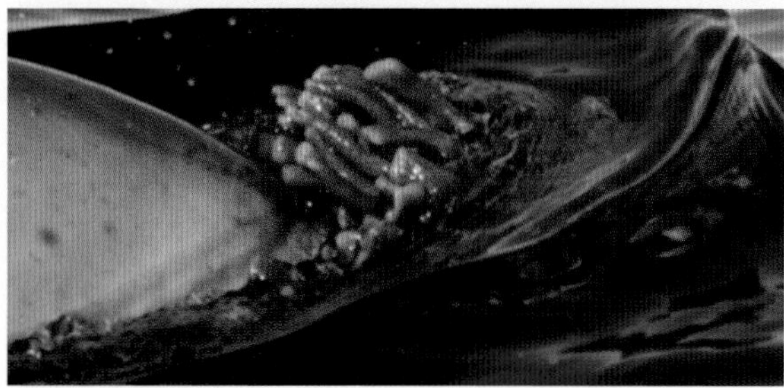

1 THE HYPOTHESIS

We hypothesized that spongers target prey that are difficult to detect with echolocation, meaning they must probe the substrate to find them, and as a result, use sponge tools for protection since the substrate where sponging occurs is littered with sharp rocks and shells.

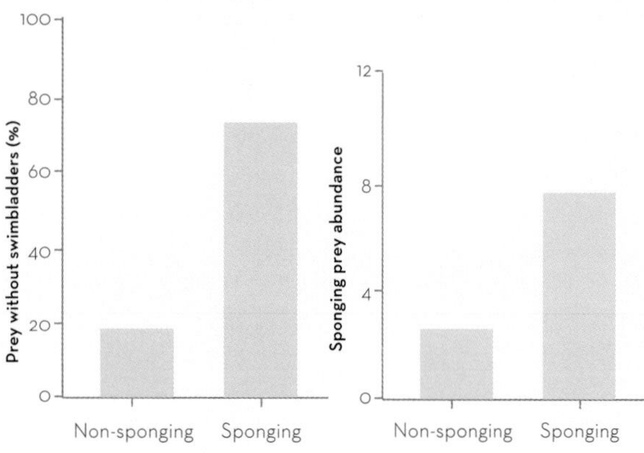

2 THE RATIONALE

Given this hypothesis, which prey would be hard to detect without echolocation? Well most fish have a swimbladder, a gas-filled organ that controls buoyancy. Swimbladders have a drastically different density than water, allowing dolphins to detect fish using echolocation. However, over time some bottom-dwelling species have lost their swimbladders since they have little need to go up and down in the water column. It is these fish without swimbladders, that we hypothesized spongers target.

3 THE TEST

To test this, we performed human sponging by pushing a pole with an attached sponge along the substrate in areas where we see dolphins sponging. This human sponging was filmed and the prey that we scared up were compared to prey that we filmed when divers just swam along the bottom without probing the seafloor. This allowed us to determine whether sponging provides access to a unique food source.

4 THE RESULT

We found that sponging (or, at least, human sponging) provides access to prey without swimbladders (see the first graph), mostly barred sandperch (just visible above). Furthermore, by sponging, dolphins have access to an increased abundance of these prey (the second graph). In comparing spongers' potential diet to that of other dolphins in the bay, we found the spongers' diet appears to be unique. This dietary difference has subsequently been confirmed through DNA and blubber analysis.

Above The sponge mostly used by some of the dolphins in Shark Bay is *Echinodictyum* sp., (top), followed by *Ircinia* and *Pseudoceratina* sp., the two below it. We have only ever seen spongers use *Axos flabelliformis*, the fourth species shown here, once or twice.

FACTORS INFLUENCING CETACEAN TOOL USE

As might be expected, cetacean tool use is rare. Cetacean flippers, flukes, and beaks are not well suited for manipulating objects, and the marine environment is not a very good medium for tool use. That said, cetacean tool use does tell us something about the factors that promote tool use among animals more broadly.

PHYSIOLOGY & ECOLOGY

Cetaceans, like other marine mammals and fish, have an efficient streamlined body that generally lacks appendages capable of manipulation. While cetacean underwater activity has not been observed well enough to know the extent of tool use, the properties of aquatic environments are also not conducive to tool use. For example, buoyancy counteracts gravity, meaning that potential tools are lighter, making them less useful. Furthermore, given the movement and viscosity of water, striking or even controlling objects underwater is more difficult than in air. It is perhaps not surprising then that there just aren't that many examples of cetacean tool use when compared to terrestrial animals such as primates.

Most examples of cetacean tool use fall into two general categories: the use of water as a tool (bubbles, jets, waves, mud plumes suspended in water) and the use of objects near the surface or on the substrate, where there might be more control of the object. Where water is the "tool," cetaceans use their unique physiology, which is well adapted to control water, to manipulate the surrounding medium. In the case of object use at the substrate, handless dolphins use their beak to manipulate or hold objects, much like birds use their beaks to probe for prey. Thus, despite their streamlined profiles, cetaceans have found innovative ways of manipulating their environment to achieve a particular end, usually to catch fish!

COGNITION

Like some tool use on land, the presence of tool use among cetaceans appears to require learning and higher cognition, especially when the behavior is learned from others. Like other big-brained tool users (primates, sea otters, corvids) cetaceans show substantial flexibility in whether or not they use tools and in the types of tools they use, evidence that the occurrence of tool use depends on learning.

The variation in tool use both within and between populations, suggests that these behaviors were innovated and socially learned, passed down for generations (see also Chapter 6). Humpback whales in both the northeast Pacific and the northwest Atlantic engage in bubble-net feeding, and the same individuals engage in the behavior repeatedly, suggesting that some have learned the behavior while others have not even though they use the same habitats. In contrast, only a subset of Floridian bottlenose dolphins, those that use shallow sand banks, engage in mud-ring feeding, suggesting that this is a habitat-specific tactic.

Perhaps the best case of both individual and social learning in cetacean tool use comes from the spongers of Shark Bay. First, although many dolphins inhabit the channels where sponging occurs, and even associate with spongers during sponging, only a small fraction become spongers. Those dolphins that do adopt sponging learn it from their mothers and almost exclusively forage using sponge tools for the rest of their lives. This suggests a sensitive period when foraging tactics are socially learned.

The sex bias adoption of sponging is another indicator of social learning. More than 90 percent of daughters born to spongers become spongers,

Opposite top A sponge-foraging dolphin and her calf in Shark Bay, Western Australia. Most females calves born to sponging dolphins pick up the behavior from their mothers and then almost exclusively sponge forage for the rest of their lives.

Opposite bottom One of the few male spongers named Kooks. All of his sisters and mother are spongers. Kooks is in his early teens and has been sponging since he was a calf. Researchers are still waiting to see if he will continue to sponge often when he joins a male alliance. Kooks is wearing an *Ircinia* sponge.

compared to only 50 percent of sons. Finally, spongers show individual learning in that they continue to improve in their technique over the years in a variety of ways, peaking in efficiency around mid-life (early twenties).

When tool use requires learning, it is often associated with social tolerance, prolonged development, and relatively large brains or elaborated cognition, features exemplified by most cetaceans. More cognitively complex tool use is generally characterized by tool manufacture, tool composites (combined objects), tool reuse, cultural transmission, and cumulative technologies, several of which are present in cetaceans. Humpback whales and bottlenose dolphins appear to carefully construct (manufacture) their bubble nets and mud rings respectively. Spongers often reuse their tools over and over again, for up to several hours. Cultural (social) transmission is also evident in some cetacean tool use, particularly for sponging.

Above Two juvenile female spongers, Ashton and Osprey, bowriding our boat with sponges on. This image is rare because spongers rarely associate while they have their sponges on, but the lure of a bowride meant that they interrupted their sponging to come for a quick spin. Although spongers preferentially associate with each other, they do not usually wear their sponges when doing so.

SOCIALITY

In some terrestrial species, tool use seems to influence and/or be influenced by the social environment. In cetaceans, we really only have detailed evidence of this in sponging dolphins. Sponging is vertically socially learned, from parent to offspring. This is perhaps not surprising, since in marine environments naïve individuals would rarely observe others besides their mothers using tools and/or encounter tool products aside from those of their mothers. Water currents likely wash away tool artifacts, and even those heavy enough to remain are eventually buried by

Vertical transmission of learned behavior

More than 90 percent of daughters compared with 50 percent of sons of spongers learn from their mother how to sponge, but because sons do not pass on the learned behavior to their own progeny, it becomes a dead end through the male line.

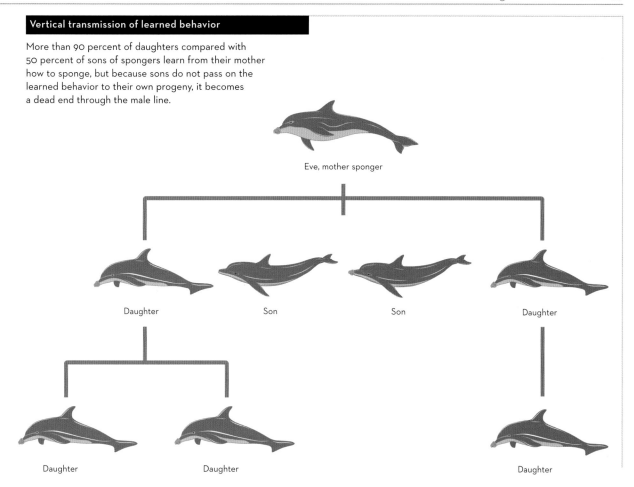

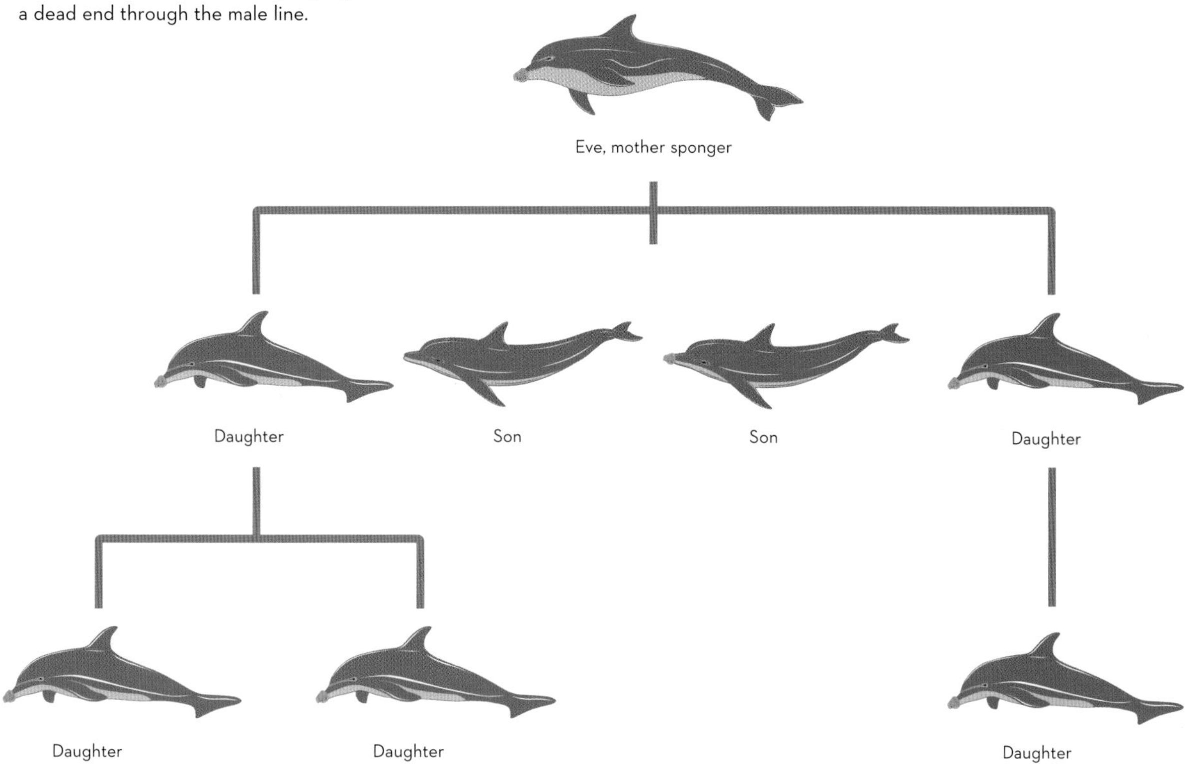

Eve, mother sponger

Daughter

Son

Son

Daughter

Daughter

Daughter

Daughter

sediment. This contrasts with terrestrial tool use where individuals regularly visit tool-using sites and even use tools in groups. Both chimpanzees and capuchins crack nuts in groups at regular sites with ample stone tools, creating opportunities for individuals to learn from other individuals as well as their mothers.

However, the vertically learning of sponging is not equal among the sexes. Sponging emerges in the second or third year of a bottlenose dolphin's life, so both sexes have exposure to thousands of hours of their mothers' sponging. Yet half of sons do not become spongers, perhaps because it too costly for males in terms of reproductive success. As described in Chapter 5, in adulthood, males form alliances with other males, which are critical for gaining access to fertile females. Sponging, which occurs only in deep channels, may constrain a male's ability to consort with females. For example, in the eastern gulf of Shark Bay, channels occur in less than 10 percent of our study area, and total channel area is less than half the size of a typical male home range and a little over one quarter the size of a combined male alliance range. Furthermore, it takes decades to become a fully proficient sponger and since males may not be able to sponge as often as females, they might not be very good at it.

Sponging is not only influenced by social factors, it also appears to influence a dolphin's sociality. For example, spongers preferentially associate with other spongers, even forming sponging "cliques." Thus, it appears that sponging serves an affiliative function and distinguishes between groups. This, combined with the fact that sponging is socially learned, makes sponging not only one of the a few cases of cetacean tool use, but an example of animal culture too, a topic discussed in greater detail in Chapter 6.

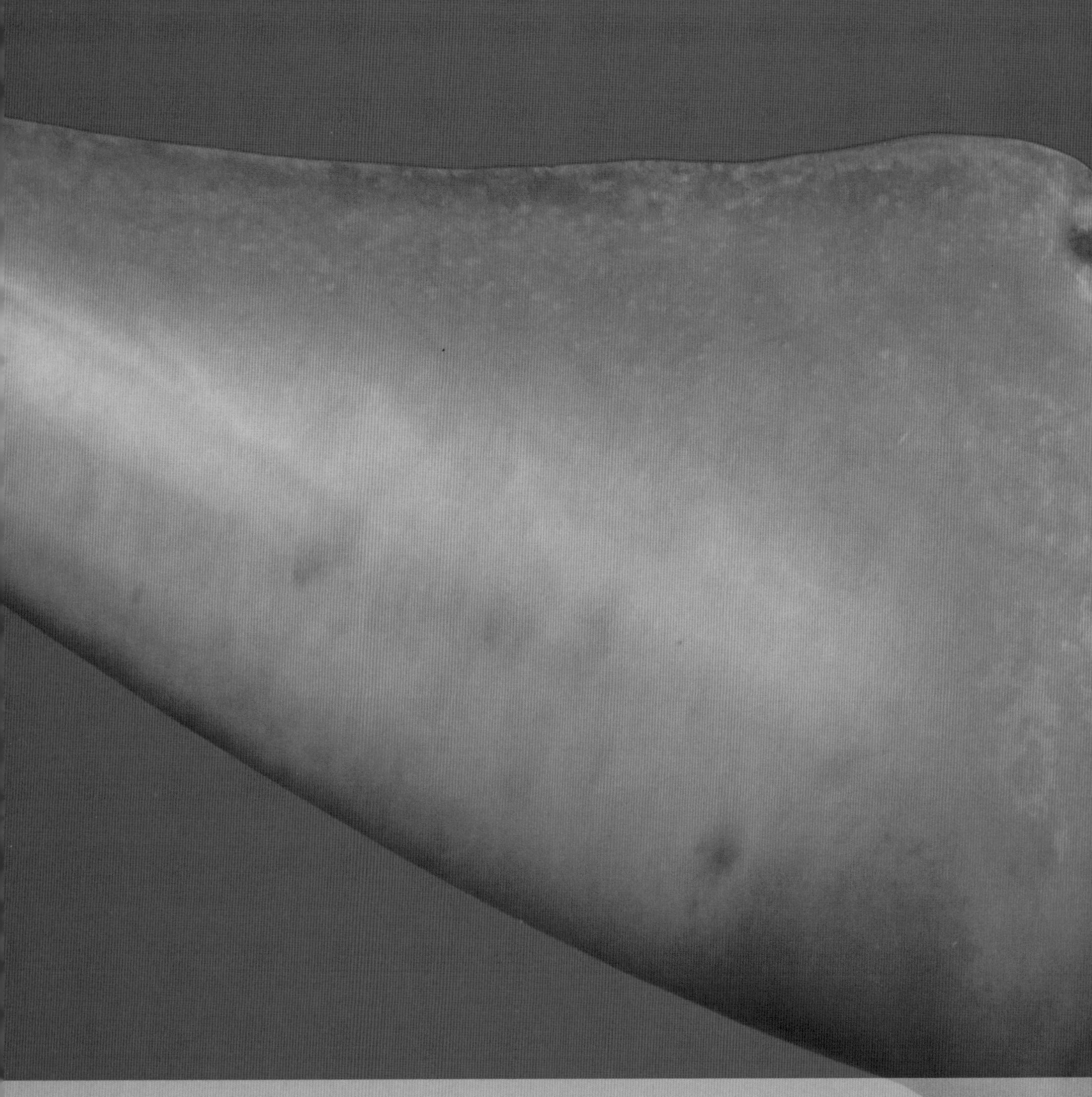

8 US & THEM

Andrew Read

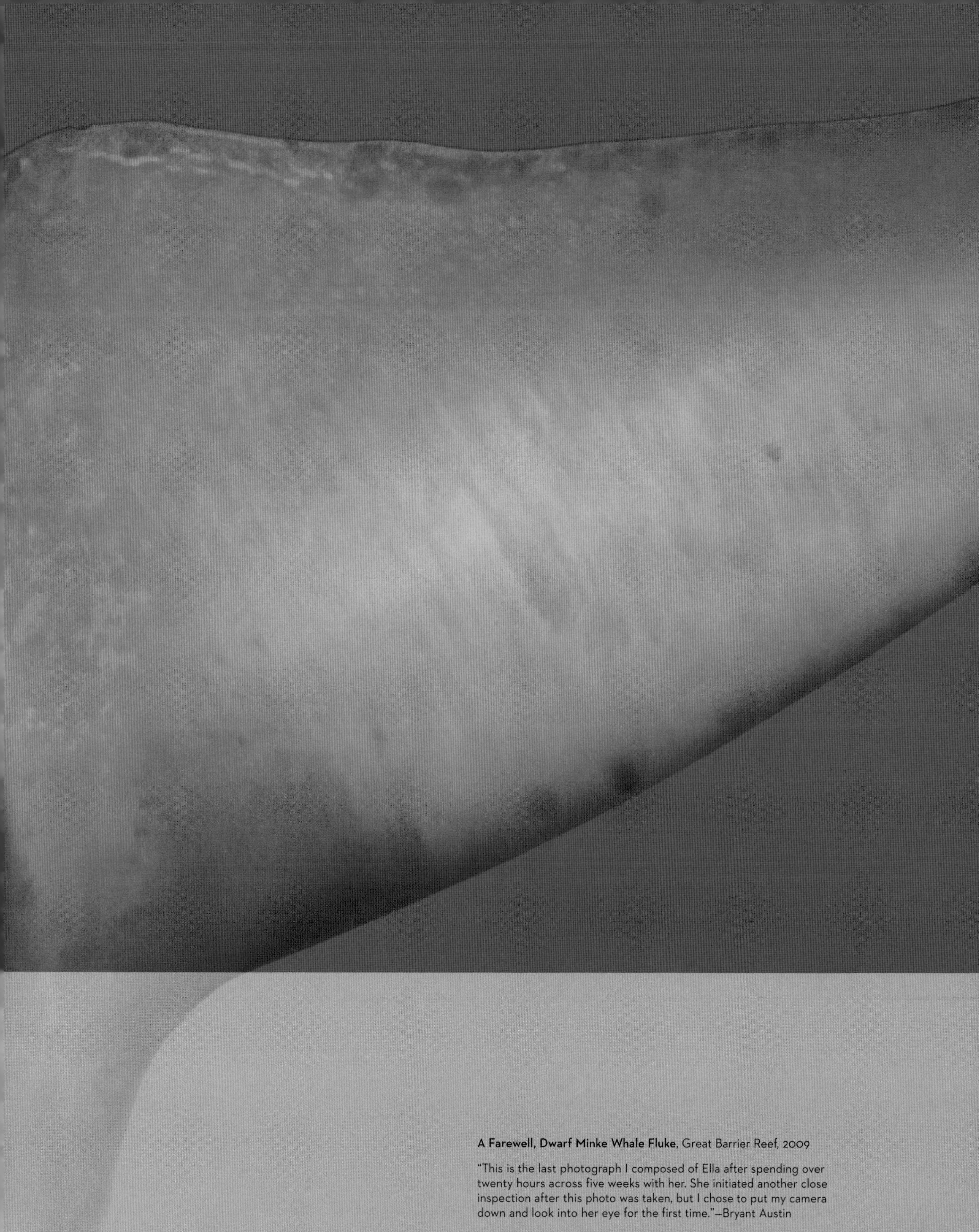

A Farewell, Dwarf Minke Whale Fluke, Great Barrier Reef, 2009

"This is the last photograph I composed of Ella after spending over twenty hours across five weeks with her. She initiated another close inspection after this photo was taken, but I chose to put my camera down and look into her eye for the first time."—Bryant Austin

THE CASE FOR CONSERVATION

Cetaceans are among the most intelligent animals, with long lives, exquisite learning ability, intensive long-term social bonds, varied cultures, and even the ability to use tools. They are beloved by many of us for these traits, but they also play important ecological roles in our oceans. From either perspective, the loss of a cetacean species diminishes our world. Populations of whales, dolphins, and porpoises are especially vulnerable to certain human activities because they are so long lived and slow to reproduce. It is difficult for any long-lived organism to sustain even moderate levels of removals, regardless of whether these removals are intentional, such as hunting, or unintentional, as in accidental death in fishing gear. And, further, because of their life histories, it takes a very long time for these populations to recover from depletion.

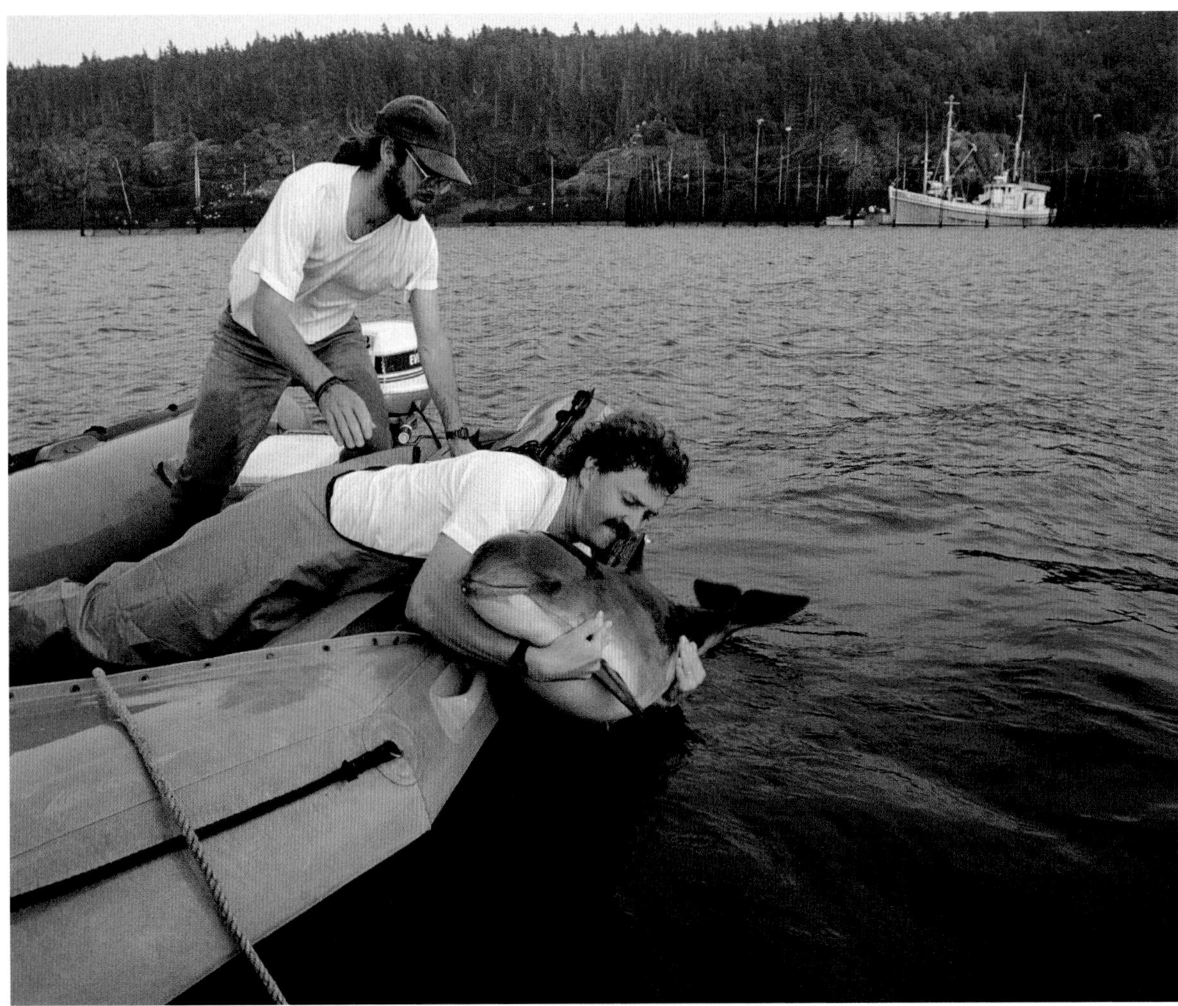

Over the past half-century there has been an explosion of interest in the conservation of these iconic animals. Put simply, conservation is the science and practice of conserving the Earth's biological diversity. So, in this chapter we are concerned with conservation of the world's whales, dolphins, and porpoises, which amount to approximately 90 species. I use the term "approximately" because the number is constantly changing as scientists collect information that suggest species should be split apart or lumped together. And, amazing as it seems, we are still discovering species, including a new species of beaked whale that was identified from a stranded specimen in Alaska as I wrote this chapter.

Conservation differs from many other branches of science in several important ways. First, and foremost, it is a normative discipline—one imbued with important values. Conservation scientists believe that biological diversity is inherently a good thing. An obvious corollary of this statement is that untimely extinctions are a bad thing. That is not to say that extinctions do not occur naturally—any cursory glance at the fossil record shows that is true—simply that we should try to minimize the number of species lost as a result of human activities.

The good news is that, unlike some other groups of animals and plants, diversity of the world's cetaceans is still relatively intact; we have lost only a single species, the baiji or Yangtze river dolphin, since we started keeping track of biological diversity a few hundred years ago. The bad news is that several species are in real trouble and at least one, the vaquita, is now teetering on the brink of extinction (see pages 172–5). And the disconcerting news is that for many species we simply don't know enough to determine status.

CONSERVATION ASSESSMENTS

The first step in practicing conservation is to assess the status of the animals we are working with. We want to devote our scarce resources to the species most in need of, and likely to benefit from, conservation interventions. Fortunately, there is a global inventory of the world's plants and animals—the Red List, maintained by the International Union for Conservation of Nature (IUCN). Each species is assessed by a group of expert scientists, including several of the authors of this book, who assess its abundance, trend in population size, and habitat. These observations are used to assign each species to a conservation category varying from the most imperiled species (Critically Endangered) to those in good shape (Least Concern). Many species are listed simply as Data Deficient, because we know too little about them to conduct a proper assessment.

The IUCN Red List

The IUCN Red List process for assessing conservation status of the world's plants and animals is the most comprehensive inventory of the conservation status of species. It adopts criteria to assess the risk of extinction of thousands of animals and plants. The Red List highlights for policy makers and the public the urgency of conservation efforts and assists the international community involved in the work to minimize the risk of species extinction.

source: IUCN

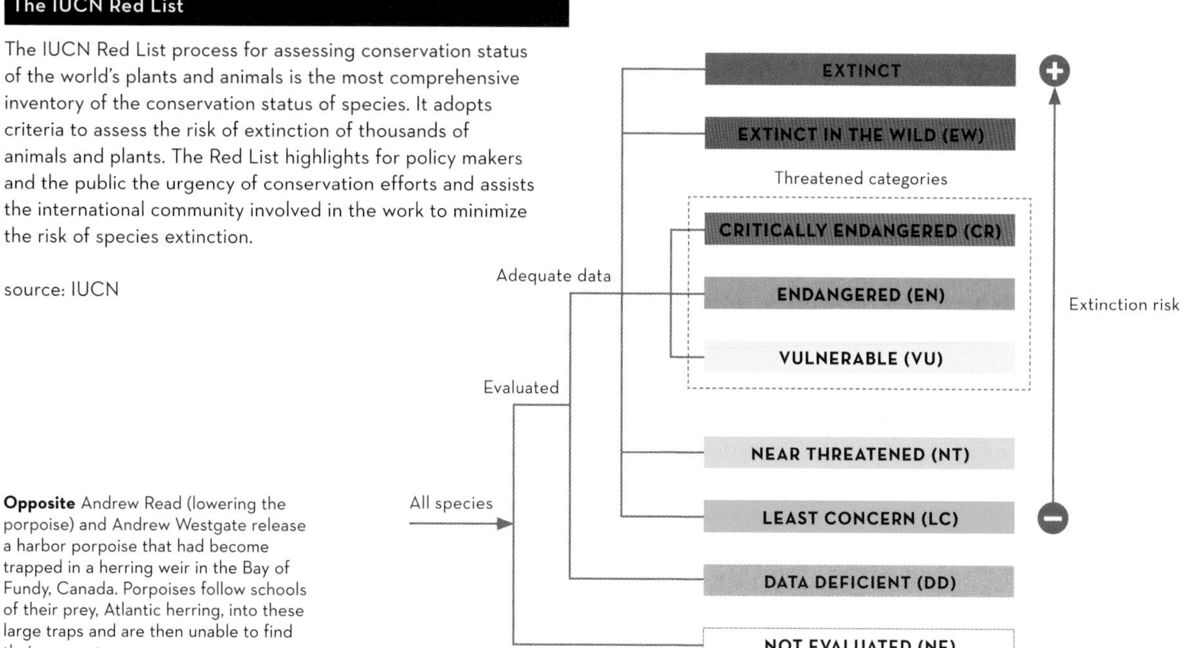

Opposite Andrew Read (lowering the porpoise) and Andrew Westgate release a harbor porpoise that had become trapped in a herring weir in the Bay of Fundy, Canada. Porpoises follow schools of their prey, Atlantic herring, into these large traps and are then unable to find their way out.

Above Ganges river dolphins, or susus, inhabit one of the world's largest river systems. Their habitat is fragmented and degraded by dams and barrages and much of their aquatic range is lost to irrigation during the dry season.

The eight most endangered cetaceans

CR = Critically Endangered EN = Endangered source: IUCN

SPECIES	IUCN STATUS	CAUSE OF ENDANGERMENT
Vaquita	CR	Fisheries
Sei whale	EN	Hunting
Blue whale	EN	Hunting
Fin whale	EN	Hunting
North Pacific right whale	EN	Hunting
Hector's dolphin	EN	Fisheries
North Atlantic right whale	EN	Fisheries
Ganges river dolphin	EN	Fisheries and habitat

Above The vaquita is the world's most endangered cetacean, with only a handful of individuals remaining. Vaquitas are killed by poachers in illegal nets set to catch a large sea bass, the totoaba.

One cetacean, the baiji, is known to have been lost in recent human history. Another, the vaquita, is listed as Critically Endangered (CR) by the IUCN, and seven other species are listed as Endangered (EN). These eight cetaceans, listed in the table opposite, are in greatest need of conservation attention. When sufficient information exists, we also try to assess the status of the populations that comprise a species. In some cases, a species is not in danger of extinction, but one or more of its populations are endangered. For example, the killer whale is not endangered at a global level, but the southern resident population inhabiting the waters of Puget Sound off the state of Washington is considered Endangered. You can look up the status of all the world's cetaceans on the IUCN Red List web site.

The next step, and sometimes a surprisingly difficult one, is to understand what is causing a species or population to be endangered. In some cases the cause of endangerment is clear—for example, too many animals are being killed in a fishery. In other cases, all we know is that the population is small and/or declining and the nature of threat needs to be determined. Once we understand the cause of the decline, the really hard work begins.

Mitigating the threats to these endangered cetaceans involves working closely with human communities. This is because the dynamics of these endangered species are inextricably linked to the human activities that threaten them. For example, as described below, saving a species threatened by fisheries will require a change to fishing practices. And, as you might imagine, it is difficult to persuade a fisherman to change the way he or she fishes, particularly if such a change incurs a cost. As a consequence, much conservation work involves negotiation and involves economists, social scientists, and, in the United States, lawyers.

As we look over the conservation status of the world's cetaceans, we find good and bad news. Of the eight species listed as Endangered or Critically Endangered by the IUCN, three (the blue, fin, and sei whales) are now recovering from past overhunting. Others, such as the North Pacific right whale, were depleted by whaling but show no signs of recovery. And, the most challenging cases are those involving accidental deaths in fishing gear. There is more about these species in the following pages, where I review the most important threats to the world's cetaceans and highlight some of the ways that we are working to conserve these species.

HUNTING

Whales and dolphins fascinate us, but have also provided humans with an important source of food for thousands of years. Large-scale hunting of cetaceans is now a thing of the past, although some relatively small hunts still persist today. The decimation of whale populations by commercial whaling fleets in the twentieth century remains a sobering reminder of the vulnerability of these animals in the face of modern technology.

COMMERCIAL WHALING

The era of industrial whaling, which began at the start of the twentieth century and lasted almost a hundred years, operated at a scale that is difficult to imagine. Quite literally, millions of whales were killed during this period—a statement which is both remarkable and terribly sad. Most of these whales were killed for their blubber, which was rendered into oil and turned into margarine. Given the scale of this persecution, it is quite remarkable that no species of cetacean was driven extinct by hunting.

When Svend Foyn, a Norwegian whaler, invented the grenade-tipped harpoon in 1870, he had no idea of the havoc he was about to wreak on the world's whales. Almost three million whales were killed between 1900 and 1999; two million alone in the southern hemisphere. It is impossible to imagine the ecological effects of removing such an enormous number of predators from the world's oceans.

The first species of whale to be hunted in the Antarctic was the humpback, because it was easy to find and catch. Whalers then turned their attention to

the largest and most valuable species, the blue whale. As its numbers declined, whalers targeted the next largest species, the fin whale. And so on and on, until only the smallest minke whales remained.

Approximately a quarter of a million blue whales lived in the Southern Ocean before industrial whaling. Whalers removed 99 percent of these whales and, by the time they were fully protected in 1972, perhaps only 400 remained. The decline of the species was worsened by illegal hunting by Soviet whalers after the species was protected by the International Whaling Commission (IWC) in 1966. In 1997 the population had recovered to 2,300 whales (one-hundredth of its original size) and is now growing at 8 percent per year. But even at this rate, it will take more than 200 years for the population to recover to its original size, if it ever does.

Opposite Some species, such as right whales, were overharvested in the nineteenth century and have failed to recover, despite more than a half century of legal protection.

Below A Siberian Yupik Eskimo subsistence whale-hunter throws a harpoon at a bowhead as it surfaces off St. Lawrence Island in the Bering Sea.

MODERN PROTECTION

In many areas, cultural attitudes toward whales and dolphins changed in dramatic ways during the last century. In many countries, whales were previously viewed as resources to be harvested, but are now offered full protection, with legal prohibitions on their capture or harassment. In the United States, for example, the federal Marine Mammal Protection Act prohibits the taking of any cetacean, where a take is defined as any attempt to harass, hunt, capture, or kill. This law, passed in 1972, reflects the striking changes in public attitudes toward whales, dolphins, and porpoises in the United States that occurred during the 1960s. These changes occurred elsewhere as well, and by 1982 there was enough political will within the IWC to declare a moratorium on commercial hunting, which came into effect in 1985.

Such cultural changes did not occur everywhere, of course, and several countries continue to harvest whales for commercial purposes. Minke whales are hunted in Norway and Iceland because both countries filed official objections to the IWC moratorium. Japan withdrew its objection to the moratorium under pressure from

the United States, and so cannot hunt whales legally, at least on a commercial basis. The current Japanese hunt of minke whales in the Antarctic takes advantage of a loophole in IWC regulations that allows member countries to kill whales for scientific research. It's worth noting that these regulations were written in 1946, before the development of the scientific methods described in Chapter 1. Japanese scientific whaling is really just a thinly veiled commercial harvest, in which several hundred whales are killed each year. In a case brought by the Government of Australia in 2014, the International Court of Justice concluded that it is legal under IWC regulations to kill whales for the purposes of science. Legal, perhaps, but completely unnecessary.

ABORIGINAL HUNTING

Some whales and dolphins are still hunted for subsistence, rather than commercial, purposes. The distinction between subsistence and commercial hunting lies in the fate of the products derived from the animal. In subsistence hunts, parts of the whale or dolphin are used by indigenous peoples for food. In commercial hunts, products are sold on the open market. In Alaska, for example, bowhead and beluga whales are harvested by aboriginal peoples for food. The hunt for bowhead whales is co-managed by the IWC, the US government, and the Alaska Eskimo Whaling Commission (AEWC). Each year, Alaskan whaling villages submit their anticipated need for whale meat to the AEWC and this is factored in to the number of whales allowed to be harvested. Under current regulations, no more than 67 bowhead whales can be killed each year. This small harvest allows the population to recover from the overexploitation that occurred in the nineteenth century from Yankee and European whalers. At the same time, it yields a food source that is critical for the survival of Alaskan native villages.

Outside the Arctic, true subsistence whaling is rare in the modern commercial world. But in Lamalera village in Indonesia, aboriginal whaling for sperm whales continues to this day. The whales are killed with hand-thrown harpoons and towed back to shore. The whale meat is dried and shared among members of the community. The number of whales killed each year is small and limited by the relatively simple practices (open bamboo boats and hand-thrown harpoons) employed in the hunt. Indonesia is not a member of the IWC, so this harvest falls outside normal international management practices. Nor does the government of Indonesia provide management oversight, so there has been no scientific assessments of the sustainability of the hunt, and its effects on sperm whale populations are unknown. This type

Opposite A bowhead whale killed by aboriginal subsistence hunters in Barrow, Alaska. A single whale can provide enough food to sustain an entire village community over the long Arctic winter.

Above Pilot whales killed in the Faroe Islands in the northeastern Atlantic. Entire family groups pass these remote islands annually as they migrate from spring onward toward the nutrient-rich waters of the Arctic. They are driven ashore and killed by hunters in these remote islands. The hunt, known as the Grindadráp, has been a Faroese tradition since at least the thirteenth century although the slaughter of up to 800 animals each year cannot be deemed an essential source of food for today's islanders.

of small-scale whaling is now rare, but provides a rare glimpse of how many past maritime human communities depended on whales for subsistence.

HUNTING DOLPHINS TODAY

Dolphins and small toothed whales continue to be hunted in several parts of the world, sometimes for subsistence purposes by indigenous peoples, and in other cases for profit. In some areas dolphins are killed and their meat and blubber is used as bait.

The most infamous dolphin hunt is the drive fishery that occurs in Taiji, Japan, the subject of the Oscar-winning film *The Cove*. In Taiji, schools of dolphins and small toothed whales are driven toward the shore and maintained in pens, where they are kept alive until they are butchered for their meat. A few individuals are selected for sale to captive display facilities. Many of the species hunted in Taiji are highly social species, like pilot and false killer whales, and it is difficult to conceive of the horrible fate that these animals endure. A similar fishery operates in the Faroe Islands in the North Atlantic, where family groups of long-finned pilot whales are driven ashore and killed. Regardless of one's system of cultural beliefs, such practices are clearly inhumane and should be abolished.

In other parts of the world wild dolphins are captured directly for the public display industry. The species most affected is the bottlenose dolphin, the cetacean most commonly maintained in aquaria. In most modern facilities, captive populations are sustained by in-house breeding programs and do not require supplementation from the wild. But in many substandard facilities, in which dolphin survival rates are low, deaths are offset by the addition of wild animals. This creates a never-ending cycle that perpetuates the demand for wild dolphins. In most of these live capture operations there is no assessment of the effects on wild populations.

Overall, things are certainly much better today in terms of the conservation of whales, dolphins, and porpoises than they were 50 years ago, but much work remains to ensure that all removals of cetaceans from wild populations are sustainable. And we need to bring the inhumane slaughter of dolphins and small whales in places like Taiji and the Faroe Islands to an end.

FISHERIES

Now that large-scale commercial whaling is over, the greatest threat to populations of dolphins and whales is accidental death in fishing gear, which we refer to as by-catch. Hundreds of thousands of dolphins and porpoises are killed in fishing nets each year; this mortality threatens several species with extinction. Most conservation work addressing this problem has occurred in Europe, North America, and Australia, where resources are available to conduct research and experiment with solutions. But many of the most pressing conservation issues occur in fisheries of less developed countries, where such resources are lacking.

LONGLINES—THE COST OF A FREE LUNCH

Some species of highly social odontocetes have learned to eat fish captured in longline fisheries—which seems like a free lunch. Longline fishing involves setting thousands of baited hooks suspended by branch lines from a mainline that can be up to 40 miles (65 km) or more in length. The gear is typically set in the evening and retrieved in the morning. In Hawaii, false killer whales eat hooked tuna from pelagic longlines, leaving only the heads of the fish. When the whales discover a longline, they work along the mainline and consume every hooked tuna, each of which may be worth hundreds or thousands of dollars. For the whales, this makes sense—it is easier to eat a hooked tuna than to try to catch one of these fast-swimming fish in the open ocean.

Unfortunately, the lunch is not really free. Some whales become hooked while feeding on the tuna and, even if they manage to break away, they are susceptible to infection from the hooks in their mouths and entanglement from the trailing line.

Several species of toothed whales engage in this behavior, which we call depredation. In Alaska, sperm whales exploit a longline fishery to depredate black cod, a deepwater fish. The whales listen for the sounds of the hydraulic gear used to retrieve the longlines and then, in a wonderfully delicate behavior, which the fishermen refer to as flossing, they place a branch line in their mouths and work down the line until they find the hooked fish. This saves the whales from diving several hundred yards to capture free-swimming black cod.

Efforts to keep sperm, false killer, and other toothed whales away from longline gear have all failed. Like bears that have learned to feed at a refuse dump, these animals are highly motivated to take advantage of a seemingly free lunch. As a result, this behavior has spread rapidly throughout the world's oceans, another example of the cultural transmission of a learned behavior in social odontocetes. Current attempts to minimize the effects of depredation are focusing on reducing the strength of the hooks, so that if a whale becomes entangled the hook will straighten and set it free.

GILL NETS—WALLS OF DEATH

The most dangerous type of fishing gear to small cetaceans is a gill net. These panels of nets can be anchored on the bottom or drifted at the surface. The United Nations prohibited the use of drift nets on the high seas (areas of the ocean beyond national jurisdiction) in 1992, due to large by-catches of dolphins, porpoises, and sea birds. The air-breathing cetaceans become entangled in the nets and, if they cannot surface to breathe, they will drown.

Gill nets (so-called because they trap fish by their gills) are inexpensive, easy to use, and can be deployed from small boats without electronic gear. As a result, they are one of the most common types of fishing gear used in the developing world. Several hundred thousands of dolphins, porpoises, and small whales are killed in gill nets each year. This is the largest single source of human-caused mortality for cetaceans today.

Why do dolphins and porpoises, which possess such sophisticated systems of echolocation, become entangled and die in gill nets? Perhaps entanglement is like a car accident, in which the animals are not paying sufficient attention to their surroundings. Or perhaps the animals are distracted by other dolphins or prey items in their environment, much like a human texting while driving.

Pelagic longline

Pelagic longline fishing gear showing the mainline and branch lines, each terminating in a baited hook.

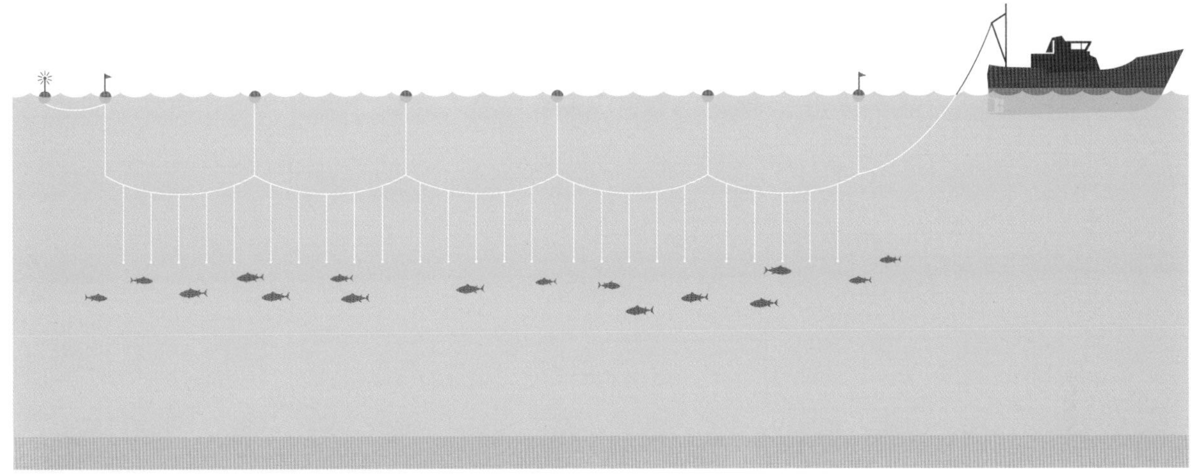

Gill net

Gill-net fishing gear showing the weighted panel anchored on the sea bed.

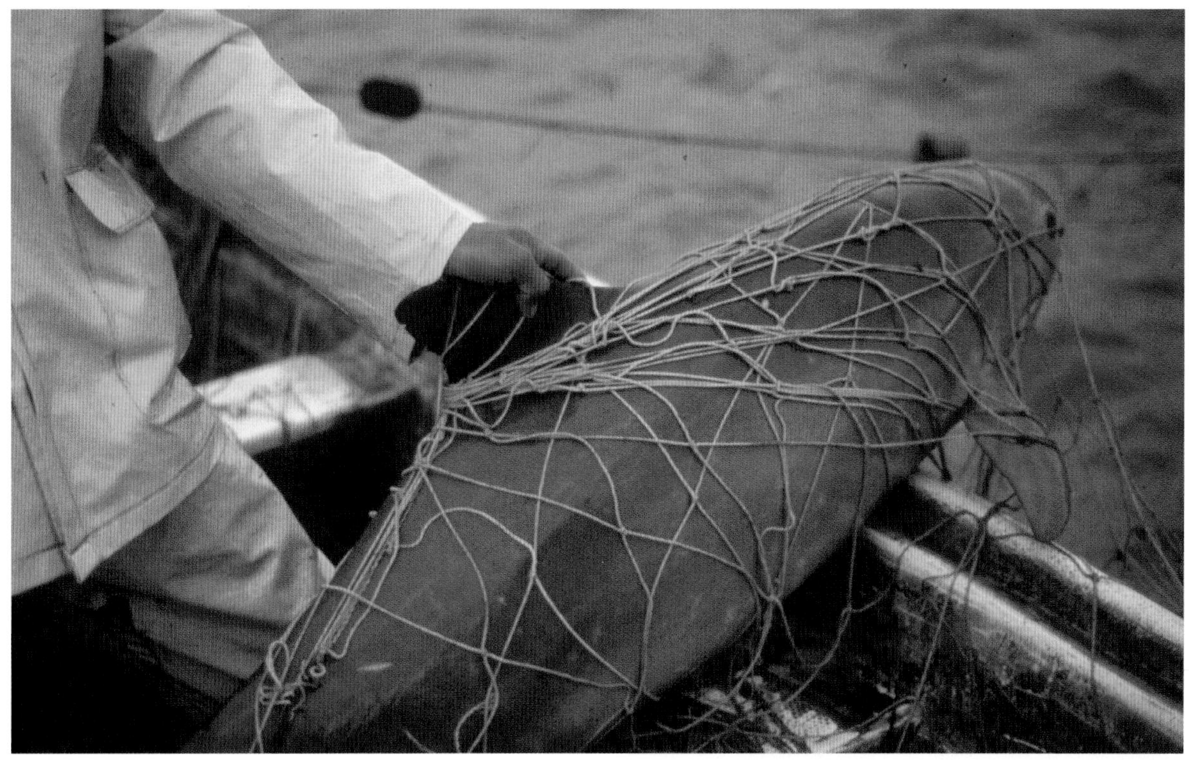

PINGERS

To alert dolphins and porpoises to the presence of a gill net, small sound emitters, known as pingers, can be placed on the nets. Pingers act to warn the animals of the net, like a flashing red light at a stop sign. In scientific experiments, pingers have reduced the mortality of harbor porpoises in gill net fisheries by up to 90 percent. Unfortunately, pingers are not as effective in the real world, as fishermen fail to replace broken devices or ensure that their batteries are charged. When used correctly, pingers can reduce by-catches of some species to sustainable levels. Unfortunately, the devices are too expensive to use in fisheries of developing countries.

Top A vaquita killed in a gill net set for totoaba in the Upper Gulf of California, Mexico.

Above Totoabas are captured for their swim bladders, which are dried and smuggled to China. The rest of the fish is discarded.

THE MOST ENDANGERED CETACEAN

A grainy black and white video shows aerial footage of four fishermen tending their gill net from a small boat. The men look up, quickly drop the net, start their boat, and race off into the distance. The men are poachers, using gill nets to catch totoaba, an endangered sea bass found only in the Gulf of California, Mexico. The video was made by the conservation group Sea Shepherd, using night-vision cameras operated from an aerial drone.

Totoaba are giant fish that weigh up to 220 lb (100 kg) and can live for 30 years. Their flesh is delicious, but their value to the poachers lies in their swimbladders. The bladders are dried and smuggled to China, where they are used in folk medicines. Totoaba are an endangered species of sea bass, so their harvest is illegal. But a single swimbladder from a large totoaba

Pingers—acoustic alarms on nets

The sound emitted by pingers deters dolphins and porpoises from swimming into gill nets and offers a partial solution to the problem of cetaceans becoming trapped in fishing gear.

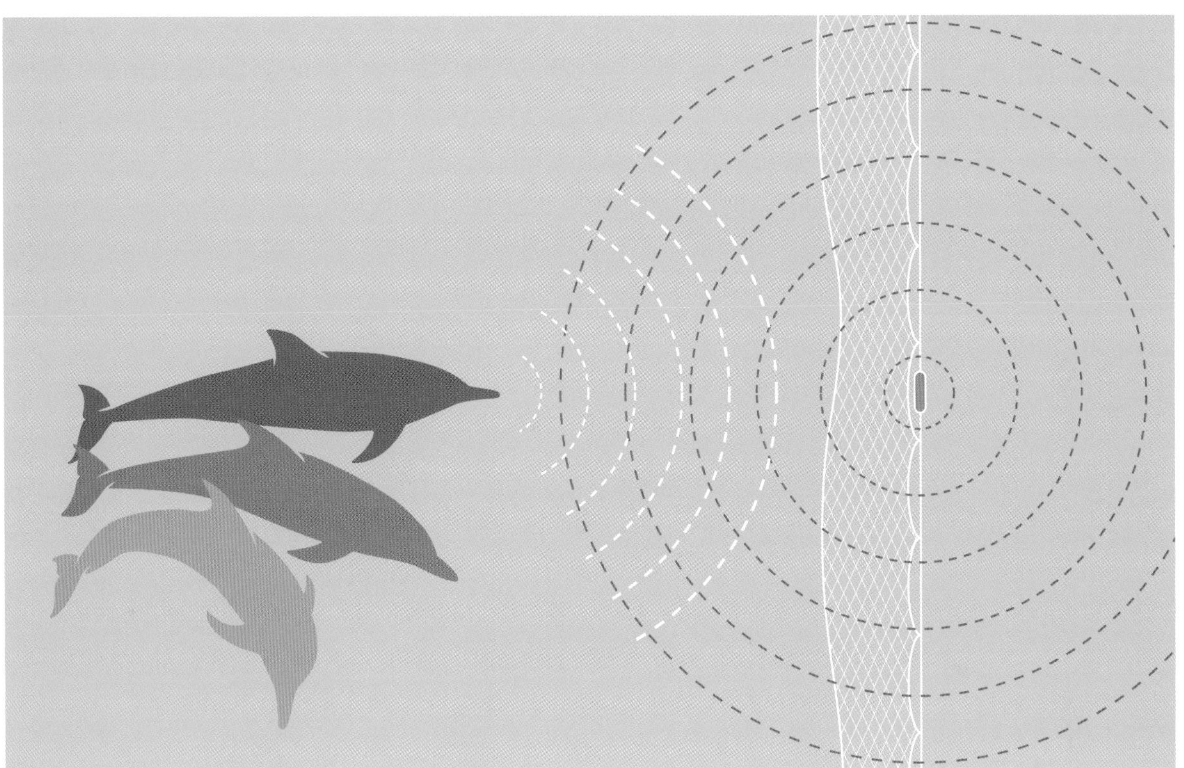

can fetch US$4,000-5,000 on the black market, so there is an enormous financial incentive to catch the giant fish. Dried totoaba swimbladders are sometimes referred to as aquatic cocaine—one recent seizure of 86 lb (39 kg) had a street value of US$750,000 in Hong Kong.

This illegal fishery targets the totoaba, but also kills the most endangered cetacean. The vaquita, also known as the desert porpoise, is found only in the Gulf of California. It is the smallest cetacean, reaching only 5 ft (1.5 m) in length, about the same size as a large totoaba. Vaquitas drown in the illegal gill nets that poachers use to catch totoaba.

A survey conducted by Mexican and American scientists in 2015 revealed that only about 60 vaquitas remain. And, in March 2016, Sea Shepherd recovered the bodies of three vaquitas killed in totoaba nets. The Government of Mexico is working hard to prevent the extinction of the desert porpoise—President Enrique Peña Nieto deployed the Mexican Navy to patrol the Upper Gulf of California to prevent poachers from setting their nets. But the poachers set their nets at night and keep tabs on where the Navy patrol boats are operating.

The world's smallest porpoise is hurtling toward extinction, tangled in a web of poachers and organized crime, propelled by an insatiable demand for totoaba swimbladders in China. Can the Mexican Navy and Sea Shepherd stop the totoaba poaching? Current conservation efforts are focused on capturing the last few vaquitas and bringing them into a sanctuary in the Upper Gulf of California. Vaquitas have never before been captured or kept in human care, so this is a high-risk proposition. At the time of writing, there seems to be no other choice for the species.

CASE STUDY: **USING PASSIVE ACOUSTIC MONITORING TO COUNT VAQUITAS**

The traditional types of surveys we use to count dolphins and whales employ visual observers working from boats and using standardized approaches to counting animals. When these surveys occur over big areas, large research vessels and multiple observers are needed, making these surveys very expensive. In cases where we need to determine population trend, it may be too expensive to conduct annual surveys. In such cases, another approach is needed.

1 USING SOUND TO DETECT VAQUITAS

In the Upper Gulf of California, Armando Jaramillo-Legoretta and his colleagues developed a clever approach to monitor the abundance of the vaquita. Like most porpoises, vaquitas emit narrow band high-frequency clicks that they use for echolocation and communication. These clicks are produced frequently enough that it is possible to use them to track the presence of the animals. To determine the relative abundance of vaquitas, Jaramillo deployed an array of 46 click detectors in the refuge that encompasses the range of this endangered species. The detectors are anchored to the sea floor and left in place for several months. These innovative devices, known as C-PODs, log the occurrence of each vaquita click.

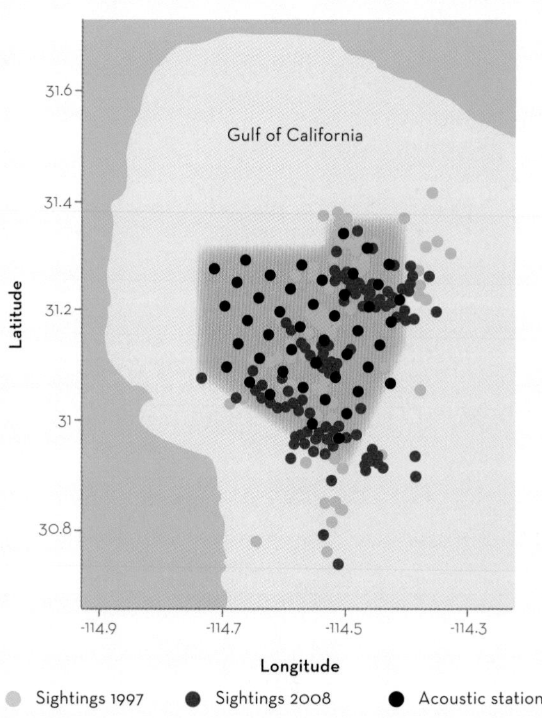

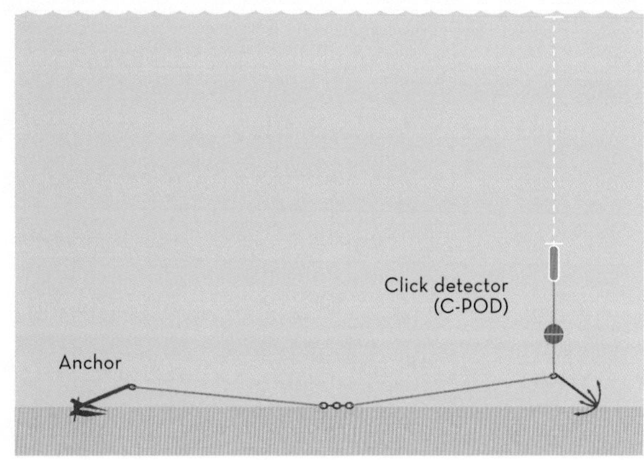

Above The survey area monitored to assess the abundance of the vaquita in the productive, murky waters of the upper Gulf of California that separates the Baja California Peninsula from the Mexican mainland.

Left The very high levels of illegal fishing in the range of the vaquita mean scientists must deploy C-PODs without any surface floats. Poachers will remove any C-PODs marked by surface floats.

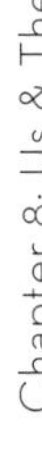

Left Without more effective intervention, the vaquita will soon be extinct. The desert porpoise is found nowhere else in the world—its demise would be a tragic loss to cetacean diversity.

Below left Armando Jaramillo and Gustavo Cárdenas from the Mexican environment ministry (SEMARNAT) prepare to launch a C-POD. A grid of C-PODs records the very high-frequency echolocation clicks that vaquitas use to find fish and also enables scientists to track their numbers.

2 FROM CLICKS TO NUMBERS

The C-PODs are retrieved at the end of the field season and the data are downloaded and examined. Jaramillo uses the number of clicks per day as a measure of the relative abundance of the vaquita. He assumes that, as the number of vaquitas rises and falls, so will the number of clicks they produce. Every ten years or so a true survey is conducted, using a large research vessel, so that we can generate an actual estimate of abundance. This acoustic monitoring program provides a critical means of determining whether the vaquita population is increasing or decreasing from year to year.

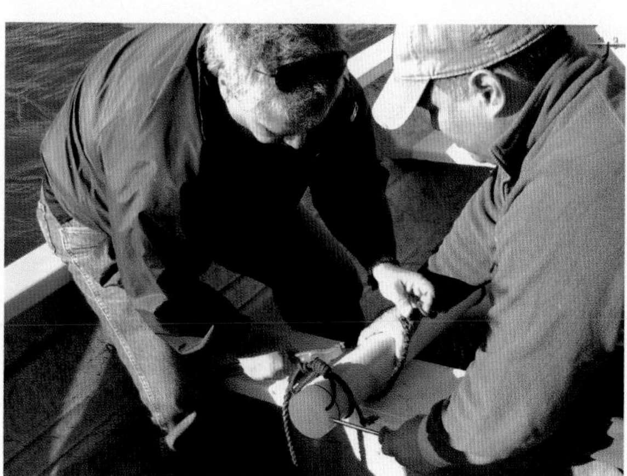

3 TRACKING THE DECLINE TOWARD EXTINCTION

The acoustic monitoring program has tracked the precipitous decline in vaquita abundance and provided the first evidence of the effects of totoaba poachers. The number of clicks per day decreased by 80 percent between 2011 and 2015, reflecting the large number of vaquitas killed in illegal totoaba nets. As we move forward, Jaramillo's program will be critical to understanding the success or failure of conservation efforts in the upper Gulf of California.

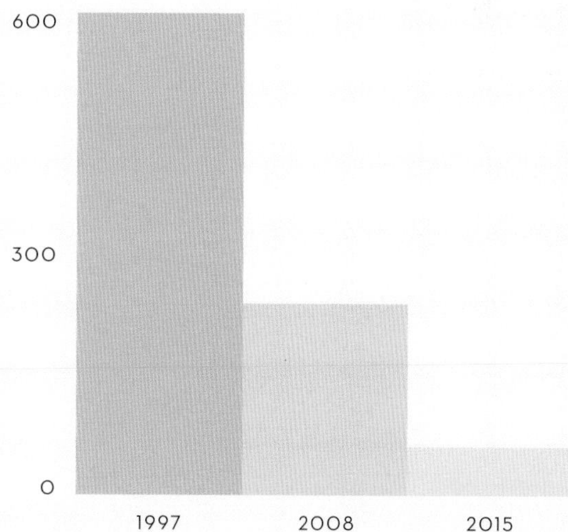

POLLUTANTS

Many cetaceans live in areas that we use intensively for commerce, recreation, fisheries, and the extraction of oil and gas. These waters also receive waste from our activities on land, including toxic pollutants that may persist for decades in the marine environment. As long-lived predators, whales and dolphins are particularly vulnerable to the bioaccumulation of pollutants that work their way up food chains to become concentrated at toxic levels in their tissues. And, like most marine organisms, cetaceans are vulnerable to the effects of environmental catastrophes, such as the Deepwater Horizon oil spill in the Gulf of Mexico.

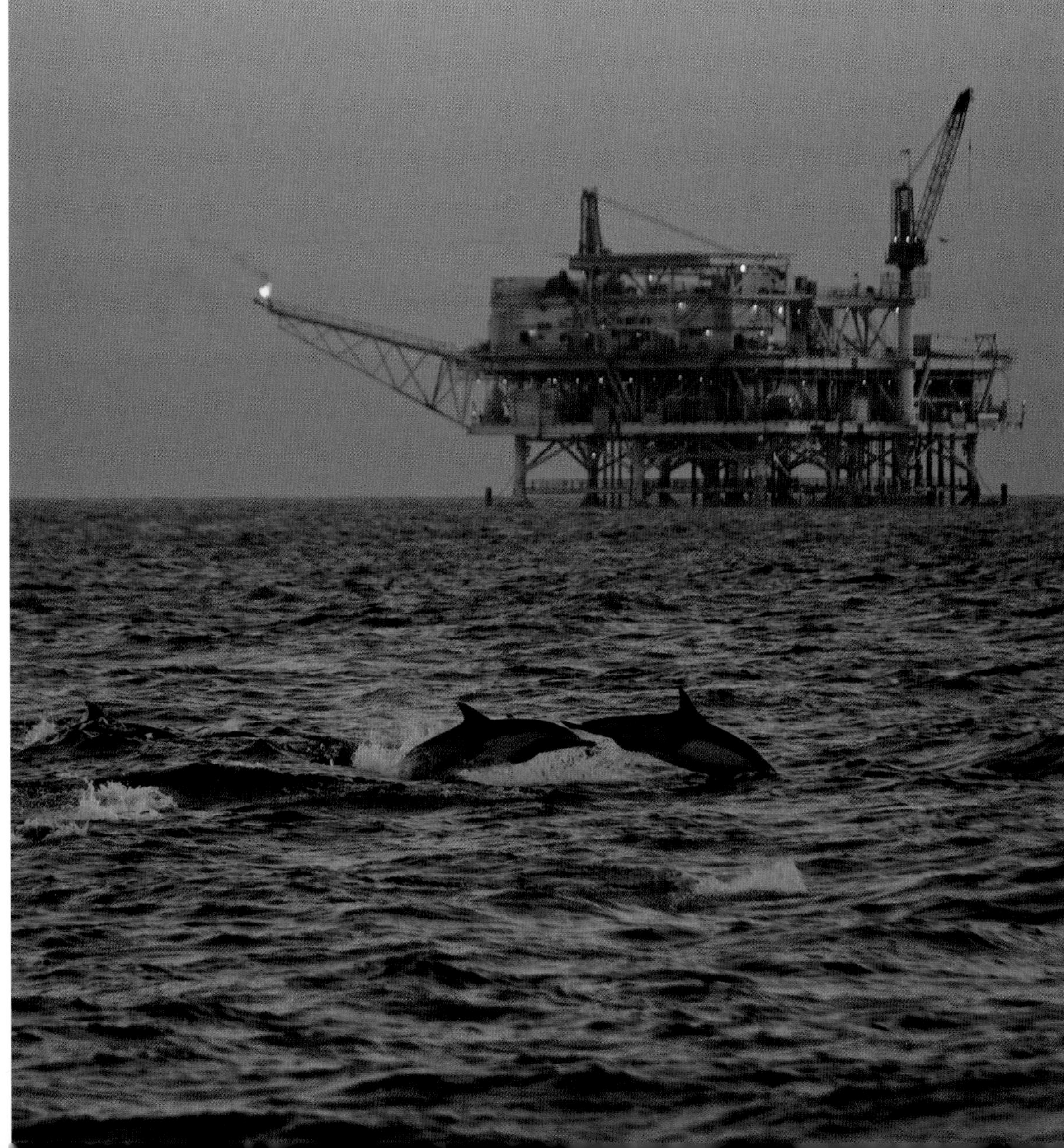

PERSISTENT ORGANOCHLORINE POLLUTANTS

Persistent organochlorine pollutants, or POPs, include some well-known chemicals, such as DDT, dioxin, and PCBs, and hundreds of other, lesser-known, compounds. The chemical structure of these compounds makes them attractive for many uses, including as pesticides, coolants, solvents, and pharmaceuticals. These organic chemicals share several characteristics, including solubility in lipids and long environmental half-lives. Some POPs, such as dioxin, are well-known carcinogens. Others inhibit reproduction, disrupt the endocrine system, or function as neurotoxins. POPs are a nasty class of materials, indeed.

The environmental effects of POPs have been recognized for more than 50 years, since the publication of Rachel Carson's landmark book, *Silent Spring*, which described the devastating effects of the insecticide DDT on populations of songbirds. DDT is no longer used widely, but its residues still occur in the tissues of many whales and dolphins. Thanks to atmospheric transport, even cetaceans in the most remote environments are not immune from these contaminants. POPs are found in the blubber of marine mammals in the Arctic at levels that pose a threat to indigenous subsistence hunters.

POPS IN SARASOTA'S BOTTLENOSE DOLPHINS

We cannot conduct experiments to determine the effects of POPs on wild populations of cetaceans, nor would it be ethical to conduct experimental trials with captive animals. Instead, our inferences regarding the effects of these pollutants come from observational studies, in which we correlate contaminant levels in tissues with some aspect of health or evolutionary fitness (e.g. survival or reproductive rates). And, as every high-school science student knows, correlation is not causation, so we must be particularly careful when interpreting the results of such studies.

Perhaps the best case study of the effects of POPs comes from the long-term research project on bottlenose dolphins in Sarasota, Florida, led by conservation biologist Randy Wells. Wells and his colleagues have followed the lives of five generations of dolphins in Sarasota Bay. The identity of each individual in this dolphin community is known, together with their maternal lineages and, increasingly, information on paternity. An important feature of Well's project is a regular health-assessment program, in which dolphins are captured, examined by veterinarians, and then released back into the wild. During these health assessments, biological samples, including blood, blubber, and milk, are taken and a suite of measurements are obtained. By linking these measures of health with subsequent observations of the dolphins after release, Wells and his colleagues can correlate contaminant levels in the tissues of each individual with important metrics such as survival and fecundity.

The Sarasota dolphins tell an important story about the divergent fate of POPs in males and females. Male dolphins accumulate POPs in their blubber as they age. In contrast, POP concentrations in the blubber of females increase until they reach sexual maturity. After females produce their first calf, POP levels drop significantly and remain relatively low (compared to males) for the rest of their lives. In a rather cruel twist, the female dolphins are passing much of their accumulated POP burden to their calves through their milk, which is lipid-rich and thus a good vehicle for pollutant transfer. The effect of this maternal transfer of POPs is particularly pernicious for the first calf produced by female dolphins. Only half of these first calves survive to weaning, much lower than the rate for subsequent offspring. Other factors, such as the inexperience of first-time mothers, may play a role, but it is clear that the transfer of POPs elevates the risk to the first calves in a significant manner. The effect of elevated POP levels on older males is unknown.

Opposite Common dolphins travel past an oil drilling platform off the coast of southern California. The introduction of pollutants from such human activities can have long-lasting environmental effects, particularly for predators, such as these dolphins, which feed at the apex of the food web.

Above White-striped dolphins swimming through a plume of oil released during the Deepwater Horizon spill into the Gulf of Mexico. We may never fully understand the extent of damage caused by this disaster on pelagic cetaceans, like these dolphins.

CATASTROPHIC OIL SPILLS

The images of the Deepwater Horizon (DWH) disaster are seared into our collective memory. After the explosion of the rig in April 2010, five million barrels of oil spilled into the Gulf of Mexico; the oil eventually coated more than a thousand miles of shoreline. Dolphins were seen surfacing in oil slicks in coastal bays and in offshore waters.

To understand the effects of this disaster on coastal bottlenose dolphins, researchers from the National Oceanic and Atmospheric Administration (NOAA), led by Lori Schwacke, conducted health assessments, using the techniques developed in Sarasota. In fact, the Sarasota dolphins were used as a control site, because DWH oil never reached the Gulf coast of Florida. Dr. Schwacke and her colleagues captured and monitored dolphins in Barataria Bay, Louisiana, an area that was extensively oiled for a long period. The dolphins in Barataria Bay exhibited a series of unusual symptoms consistent with those seen in laboratory animals exposed to oil. These symptoms included lesions of the adrenal glands and severe lung disease, likely brought about by inhaling oil in aerosol form. Strangely, many of the dolphins captured in Barataria Bay had lost many or all of their teeth.

The veterinarians involved in the health assessment concluded that many dolphins captured in Barataria Bay were extremely sick and almost one-fifth of them

Conceptual model of health effects on dolphins

After exposure to oil, health effects previously reported for mammals (the yellow boxes) could lead to abnormalities seen in Barataria Bay dolphins (blue boxes) and ultimately to decreased survival and reproduction. After Schwacke et al. (2014).

Oil exposure (ingestion)

Oil exposure (inhalation)

Adrenal effects

Lung injury

Liver injury

Maternal transfer

Hyperkalemia/ hyponatremia

Hypoglycemia

Inflammation

Anemia

Liver enzymes

Poor body condition

Survival

Disease susceptibility

Fetal death, calf survival

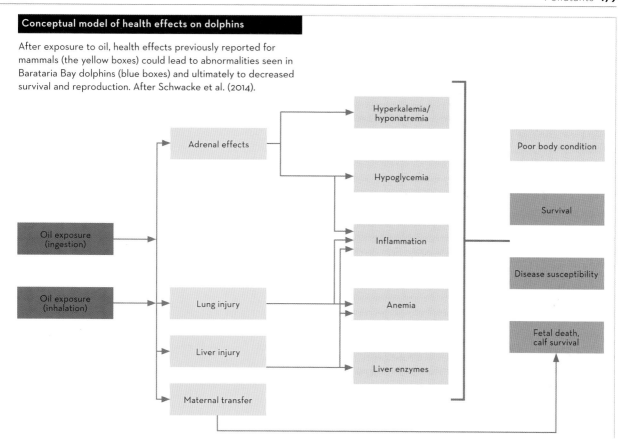

were not expected to survive. Particular concern was expressed for the fate of pregnant females. Follow-up monitoring of these females determined that only 20 percent of pregnant females produced successful calves, compared with more than 80 percent for a similar sample of dolphins in Sarasota. These reproductive losses will further impact the Barataria Bay population of dolphins.

Schwacke and her colleagues will continue to follow the health, survival, and reproduction of dolphins in Barataria Bay for many years to come, to understand the long-term effects of the DWH disaster. Lessons from elsewhere suggest that these effects will be long-lasting. In Prince William Sound off the south coast of Alaska, for example, several pods of killer whales were affected by the Exxon Valdez oil spill in 1989. The AT1 pod was particularly hard hit. Nine members of this pod died as a result of the spill and two more died shortly thereafter, reducing the pod to only 11 whales. There has been no successful reproduction since the spill and, as of 2012, the pod comprised only seven whales. The recovery of other pods has been very slow, reinforcing the idea that populations of whales need long periods to recover from such environmental disasters.

Right A bottlenose dolphin surfaces through a thick layer of oil in Barataria Bay, after the Deepwater Horizon disaster. The effects of this oil spill may last for generations.

DISTURBANCE

Human activities have the potential to disturb whales, dolphins, and porpoises, particularly in coastal waters where our presence is so ubiquitous. Sometimes the disturbance is transitory and has no long-term effects on individuals or populations. In other cases, disturbance has the potential to cause harm to individuals and populations. Such disturbances may be chronic, such as the effects of tourists swimming with spinner dolphins in Hawaii. In other cases, the disturbance may produce an acute response that places individuals in harm's way.

ARE WE LOVING WHALES TO DEATH?

As public attitudes toward cetaceans changed during the latter half of the last century, a new industry of whale and dolphin watching developed to take advantage of public interest in these animals. By some estimates, the global whale-watching industry is now worth more than US$ 2 billion annually. In the few areas where commercial whaling stubbornly holds out, such as Iceland, whaling and whale watching co-exist in an uneasy clash of old and new cultures.

Some countries have adopted regulations that apply to commercial whale-watching vessels. These rules are designed to ensure that whales and dolphins are not harmed by collisions with vessels and that the close approaches of these boats do not interfere with critical activities, such as feeding or nursing. However, in most countries there are no rules and tour operators are free to do as they please. Such unregulated ecotourism has the potential to harm individual animals and to have a population-level effect. It is a delicate balancing act to find a level of protection that allows tourists to enjoy watching these animals without causing them harm. Some of the most effective approaches involve self-policing by the tour operators themselves.

HAWAIIAN SPINNER DOLPHINS

In Hawaii, groups of spinner dolphins forage offshore at night and return to rest in coastal bays during the daytime. The dolphins choose shallow, sandy, quiet bays, where they are safe from predators, such as tiger sharks. The timing of their movements into and out of these bays is predictable. Ecotour operators have exploited the regular occurrence of spinner dolphins, particularly along the leeward sides of Oahu and Hawaii, to offer their clients the chance to swim with the dolphins. These bays are typically clear and calm, affording snorkelers an opportunity to watch spinner dolphins closely underwater. Some swimmers approach and attempt to touch the dolphins, even though such behavior violates the Marine Mammal Protection Act. Not all operators allow their clients in the water and the standards of conduct vary widely from business to business. NOAA has attempted, with little success, to establish a partnership with ecotour companies that would codify a voluntary responsible code of conduct.

When the dolphins enter the bays to rest, they begin a phase of slow, synchronous swimming in which vocal activity is very low. Resting dolphins are easily disturbed and, if disturbance reaches a certain threshold, the group will abandon the bay. The consequences of such disturbance are not easily measured, but both scientists and managers believe that, if the rest of these animals is compromised, they will be more vulnerable to predation and their nocturnal patterns of cooperative foraging will be compromised.

In the absence of an agreement among the tour operators themselves, NOAA and the State of Hawaii have been drafting regulations that will protect spinner dolphins. The current version of proposed regulations includes a rule prohibiting swimming with dolphins and preventing any person from approaching within 50 yards (45 m) of a dolphin. This "close approach rule" is similar to the regulations employed in many whale-watching areas. An alternative, and more promising, proposal is the establishment of closed areas in each bay, in which swimmers would be prohibited. All of these proposals are controversial, because of the lucrative nature of the dolphin-watching industry, and the many other uses of these coastal bays by human

Harassment of Hawaiian spinner dolphins

Hawaiian spinner dolphins feed offshore at night and then return to sheltered bays to rest during the daytime. While resting they are exposed to harassment by human swimmers.

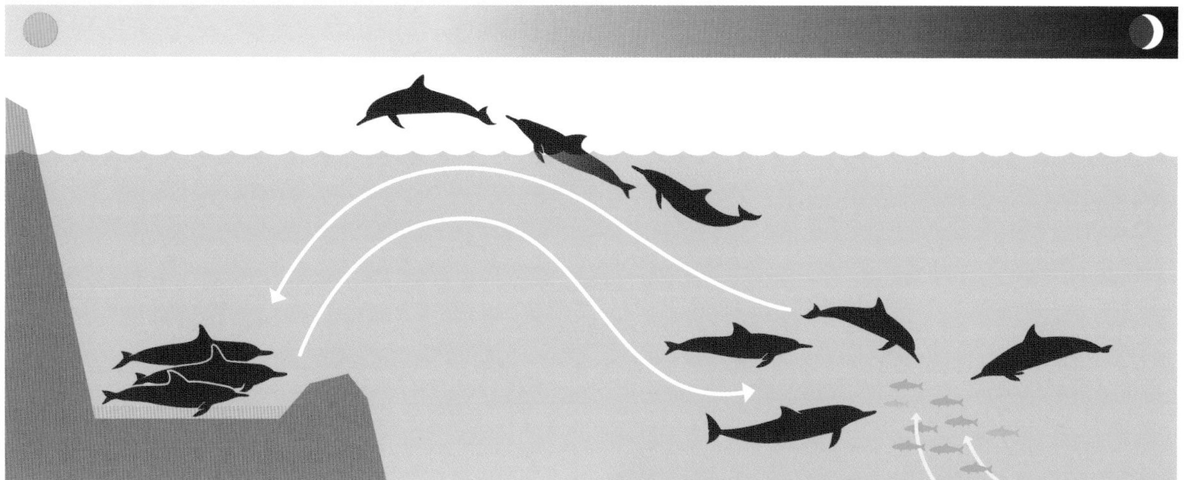

swimmers. It has taken NOAA more than a decade to issue draft regulations, underscoring the lack of agreement about how best to protect these animals.

BEAKED WHALES

In March 2000, US Navy destroyers were engaged in a routine training exercise in the Bahamas. The vessels were using tactical sonar to look for submarines as they transited over a deep canyon. Shortly after the training exercise, 17 cetaceans, including 14 beaked whales, stranded along nearby beaches. Mass strandings of beaked whales are highly unusual and this event generated an enormous degree of interest and scrutiny from the scientific and conservation communities. Post-mortems were conducted on five of the beaked whales, all of which were in good body condition, with no evidence of disease, or any traumatic injury. A report on the stranding event, jointly produced by NOAA and the Navy in December 2001, concluded that the most plausible cause of these unusual strandings was the use of tactical sonars during the training exercise.

This stranding event was not the first, nor the last, to be linked to the use of these sonars. But the conclusive link between the stranding event and the training exercise led to development of a large research program, funded largely by the US Navy, to determine how beaked whales respond to sonar. (Full disclosure: some of my research is funded by this program.) A scientific workshop convened by the US Marine Mammal Commission in April 2004 suggested that the most likely scenario was that the beaked

Above Male Cuvier's beaked whale surfacing off Cape Hatteras on the coast of North Carolina. Cuvier's beaked whales are particularly susceptible to the effects of military tactical sonars, perhaps because they perceive these signals as the sounds of predators. Under certain conditions, they may respond to the presence of sonar in a way that places them at risk of physiological harm. Sightings of this species are so rare that most scientific knowledge comes from studying stranded individuals.

Navy sonar & beaked whales

Cetaceans, including beaked whales, are exposed to may different types of sonar in coastal waters. Scientists are trying to determine what aspects of military tactical sonars cause beaked whales to strand, and under what specific circumstances.

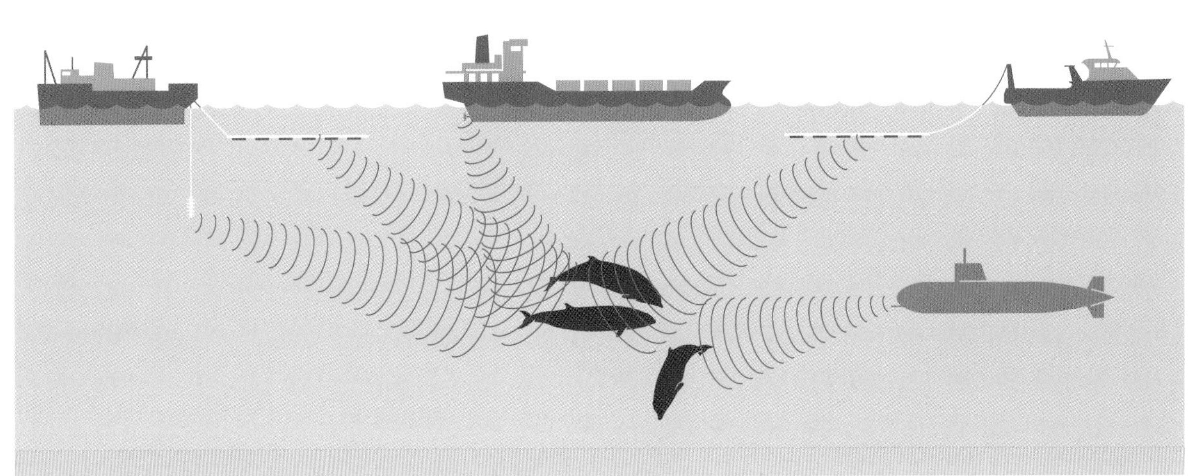

whales, the deepest mammalian divers, responded to the sounds of the tactical sonar in a way that caused them to experience decompression sickness, known to human divers as "the bends."

What would make these beaked whales change their behavior in such a dramatic fashion? Biologists Peter Tyack and Walter Zimmer noted that Navy tactical sonars shared some acoustic features with the calls of killer whales, the only predators of these deep-diving cetaceans. The cryptic behavior of beaked whales at the surface has been interpreted as an anti-predator strategy—they are vulnerable to killer whales only when they come to the surface to breathe. Perhaps the beaked whales, suddenly surrounded by the unaccustomed sounds of their predators at depth, panicked and fled in a manner that caused decompression sickness.

To understand how beaked whales respond to tactical sonars, scientists have conducted several field experiments, known as behavioural response studies. In these studies a focal whale is equipped with a digital acoustic tag (DTag). The DTag is about the size of a smartphone and is attached to the whale by four suction cups, using a carbon fiber pole. The DTag records the acoustic environment of the whale and measures the acceleration and orientation of the animal in three dimensions. This allows reconstruction of the movement of the tagged whale after the DTag is recovered—the tags float and have a small radio beacon to facilitate their recovery. The final element of the experiment is to expose the tagged whale to the sounds of a Naval tactical sonar, or a similar sound source, at levels loud enough to evoke a response, but not to cause harm.

These are complex and difficult experiments. The first step alone, attaching a DTag to a beaked whale, is a very tricky proposition. Cuvier's beaked whales, the species most susceptible to Navy sonar, can stay submerged for 75 minutes or more. And it is difficult to position the sound source in just the right place, so that the tagged whale experiences the correct received level of sound. The trials conducted to date have involved a small number of beaked whales, but their results are consistent with the hypothesis originally proposed by Tyack and Zimmer, that beaked whales respond to these sonars as if they were the sound of predators. And, perhaps unsurprisingly, the whales seem to be very adept at differentiating real sonars from the scaled sound sources used in some experiments. Much work remains to be done, but we now beginning to understand some of the factors responsible for the Bahamas stranding event.

Seismic exploration

Seismic exploration for oil and gas reserves employs some of the loudest human-generated sound sources used in the ocean. Airguns towed behind the survey vessel fire intense bursts of sound that reflect off oil and gas deposits deep in the sea floor.

SEISMIC EXPLORATION

Oil and gas companies rely on detailed three-dimensional maps of geological features below the sea floor to make decisions about where to drill wells. Geophysical exploration companies conduct surveys of the ocean floor using a series of airguns towed behind a research vessel. The airguns produce extremely high-intensity sound pulses, loud enough to penetrate miles below the sea floor. The echoes from these pulses are received on hydrophones towed behind the vessel. Airguns produce some of the loudest noises created by humans in the marine environment. Very little work has been conducted to examine the effects of these sounds on cetaceans, although we know that airguns affect the calling rate of bowhead whales and the foraging behavior of sperm whales. A fierce debate is currently raging between the seismic industry and cetacean scientists regarding the effects of airgun surveys on whale populations. Proponents of seismic exploration point to a lack of documented impacts on populations of cetaceans in areas where air guns are used frequently, such as the Gulf of Mexico. Cetacean scientists like me respond that no effects have been documented because no baseline studies were conducted prior to the onset of these surveys and, furthermore, because so few studies have been conducted on their potential effects. As I write, the debate is playing out along the Atlantic coast of the United States, where several companies have applied for permits to conduct airgun surveys.

Above Scientists are increasingly concerned about the effect of seismic surveys on cetaceans. Are these surveys responsible for the mass stranding of pilot whales and other deep-diving cetaceans?

OTHER THREATS & HOW YOU CAN HELP

We've looked at a number of the major threats to cetacean populations, but many other human activities also pose a threat, including habitat modification and climate change. Understanding these threats and working together to address their impact can make a real difference.

HABITAT MODIFICATION

Habitat modification and loss is an important issue for cetaceans that inhabit environments used intensively by humans. Foremost among these are some of the world's largest river systems, which support populations of freshwater dolphins. The Ganges river dolphin, or susu, inhabits one of the most densely populated areas of the world and its riverine environment has been disrupted by more than 50 dams and barrages. In addition to fragmenting the species into isolated units, the dams alter the river flow, reduce prey populations, and diminish the physical space available to these dolphins. Perhaps only 1,500 susus remain.

CLIMATE CHANGE

Our world is warming rapidly and we are only beginning to grasp the consequences of the enormous changes taking place as a result. The consequences of climate change are likely to be greatest for cetaceans with small, restricted ranges and barriers to their movement. Of particular concern are ice-associated species, such as bowhead whales, belugas, and narwhals of the Arctic. We cannot yet predict the effects of ice-free summers in the Arctic on these and other species. Humanity is conducting a vast, uncontrolled experiment on a global scale and we will have to do our best to document the outcome.

WHAT CAN YOU DO?

The best advice I can give you, the reader who would like to help whales, dolphins, and porpoises, is to become an engaged citizen. First, find out how your fish and seafood is sourced and avoid products captured in destructive fishing gear, such as gill nets. Next time you go to a restaurant, ask your server where your fish comes from. Consult the consumer guides produced by organizations like the Monterey Bay Aquarium, Marine Conservation Society, and Blue Ocean Institute when purchasing seafood. You can become a member of, and volunteer for one of the many environmental organizations engaged in conservation work with cetaceans. And, finally, you can become engaged directly in policy making, by writing letters supporting particular policy options. Come join us—there is much to do.

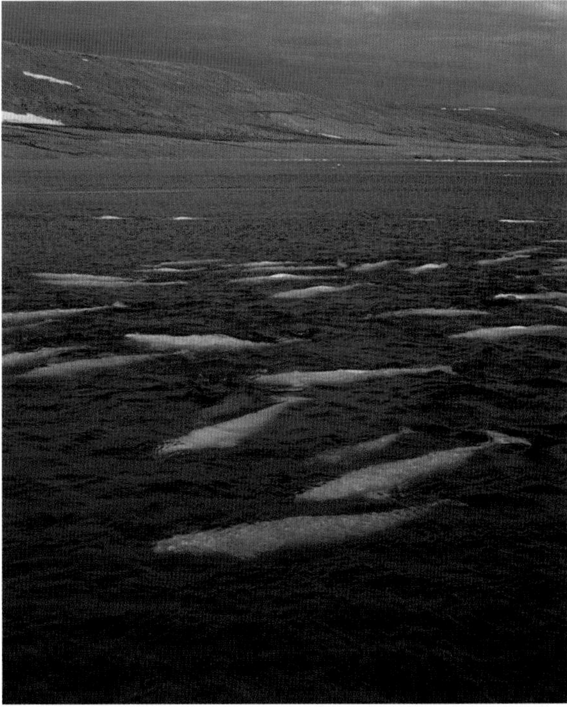

Above Even in remote environments, including Arctic estuaries, human activities have a measurable effect on populations of cetaceans. Beluga whales carry levels of pollutants in their blubber that are high enough to cause concern for the native hunters who rely on them for subsistence.

Opposite The major long-term concern for cetaceans at both poles of the earth is the rapid change occurring in their environments. How will Arctic and sub-Arctic species like narwhals and belugas adapt to an ice-free summer environment in the next few decades?

GLOSSARY

alliance A cooperative bond that persists through time, where the allied group consistently cooperates against another party. The best-documented case of alliances in cetaceans is pairs and triplets of adult bottlenose dolphin males who repeatedly cooperate to gain access to females and in battle against other male alliances.

allomaternal care Care by any individual who is not the mother of the offspring. It implies some benefit to the mother because she is released from caregiving temporarily.

aspartic acid racemization The aging technique for cetaceans and other life forms that relies on the pace at which amino acids change since the animal died. The rate of change– racemization–depends on the type of amino acid.

autonomic regulation Refers to the regulation of both the sympathetic nervous system (e.g., speeds heart rate, regulates blood pressure) and the parasympathetic nervous system (e.g., slows heart rate, increases intestinal and gland activity).

baleen Plates of keratin (epithelial tissue, the material that makes up fingernails, claws, and hair of mammals) that has replaced teeth in mysticetes. Mysticetes use the baleen to filter out water and keep prey in their mouths. Also called whalebone.

baleen whales One of two major suborders of whales that lack teeth as adults and possess baleen.

balenids Members of the Balaenidae, the family that includes right whales and bowhead whales.

beak The upper and lower jaw of a dolphin or whale. Also called rostrum.

benthic Living at or near the sea bottom.

blow Respiratory exudate or exhaled air mixed with water and lipids (fats) released by whales from the blowhole but originating from the lungs; it can appear low and bushy or tall and columnar, depending on the species, and is a useful field-identification feature.

blowhole Nostril opening at the top of the head of whales that functions in respiration; odontocetes have one blowhole and mysticetes possess two.

Blow-sampling A method for capturing blow from dolphins and whales.

bow-riding An energy-saving behavior of some odontocetes (e.g. dolphins) in which they position themselves near the bow or front of a boat to ride the pressure wave.

breaching A behavior that involves leaping partially or completely out of the water. Breaching commonly occurs during socializing so may serve a signaling function.

bubble feeding (cloud and net) Two different techniques of feeding where humpback whales create a bubble net or cloud using their blowhole that confuses or encircles prey for easy gulping.

bulla A protective bone cavity that protects the middle ear. It is particularly prominent and thick in odontocetes (toothed whales).

bycatch That portion of a catch or harvest that includes non-targeted animals. Porpoises and dolphins are vulnerable to becoming bycatch in some commercial fisheries such as gill-netting.

cetaceans Marine mammals that include whales, dolphins, and porpoises.

carousel feeding A cooperative foraging method used by killer whales and bottlenose dolphins where they encircle schools of prey (often herring) and use their tails to stun the fish.

cerebellum The hind brain that is extremely important for motor and sensorimotor learning and activity, including play. The cerebellum is quite large in odontocetes.

clicks Refers to echolocation clicks, which are broadband signals typically ranging from 30–150 kHz of short duration. Toothed whales use echolocation clicks for foraging, navigation, and social communication.

coalition Short-term cooperation between two or more individuals against a third party.

codas A patterned series of clicks used by sperm whales that are specific to their clan. These are thought to be socially learned and passed on culturally.

coefficient of association A measure of the strength of association between two individuals where 1 means that two animals spend 100 percent of their time together and 0 means they are never sighted together.

consortship Refers to an association between a female and one or more male suitors.

contact behaviors Refers to affiliative or non-affiliative contact between two or more animals. Petting is a friendly contact behavior in dolphins. Aggression, e.g. jawing and ramming, is also a form of contact behavior.

convergent evolution Refers to the fact that organisms that are not closely related in evolutionary time develop similar traits. A classic example is echolocation in bats and toothed whales.

corpus callosum A bundle of white nerve fibers that connects both brain hemispheres. In cetaceans, this is very thin, probably because of unihemispheric sleep.

cortex/neocortex Usually refers to the cerebral cortex or outer layer of the brain, including the neocortex. The outer layers are more recent, evolutionarily, and tend to have higher cognitive functions than the older parts of the brain.

C-PODs Underwater passive acoustic monitoring devices. They can remain underwater for long periods and can be used to detect and record underwater sounds, particularly marine mammal sounds.

cultures In general terms, a socially learned behavior that is transmitted between individuals and distinguishes one group of individuals from another.

delphinids Members of the dolphin family (Delphinidae) which includes 37 species, ranging from Hector's dolphin to the killer whale.

dentate gyrus Part of the hippocampus and thought to contribute to the formation of new episodic (event) memories among other functions.

demersal Living at or near the bottom of the sea above the benthic zone.

dorsal Refers to the back of an animal.

dorsal fin Located on the back (dorsal surface) of most whales, it provides a control surface to stabilize the animal and control body temperature.

DTags Digital acoustic recording tags that are usually attached via a suction cup to marine mammals and can record detailed physical and acoustic behavior underwater.

ecotourism A form of tourism where humans visiting natural areas minimize their disturbance of plants, animals, and habitat. It also refers to travel that is dedicated to conservation of the environment and welfare of local inhabitants.

ecotype Refers to a subgroup in a species (not necessarily a subspecies) that occupies a specific habitat or engages in a specific type of behavior in relation to that habitat. This is used to refer to different behaviors of killer whales that might overlap geographically, but use different habitats and behaviors within that habitat.

echolocation Production and reception of sonar (high-frequency pulses or clicks) whereby animals can scan the environment and acoustically identify objects/prey.

encephalization quotient (EQ) A measure of brain size relative to body size.

filter feeding Straining suspended matter/food particles from water via a specialized filtering structure. Mysticetes gulp water filled with prey, then push out the water through their baleen.

fission–fusion Refers to the dynamic changes in association between individuals in a population. Groups that are highly stable are low on fission-fusion. Groups that change membership frequently are high on fission-fusion dynamics.

flipper Front limbs (pectoral fins) of cetaceans that are variously shaped; for example, long and narrow in the humpback whale for rapid locomotion, and short and paddle-shaped in right whales, which aids in maneuvering and slow turns.

flipper-slapping A behavior that involves striking the surface of the water with the flippers; it likely functions in non-verbal communication.

focal sampling Entails following an individual or a closely bonded pair of individuals (e.g. mother and calf) for a pre-set period of time to record their behavior systematically and quantitatively.

frontal lobe The front portion of the brain common in mammals but there is no homolog in cetaceans.

gyri and sulci The ridges (the gyri) on the cerebral cortex surrounded by furrows or depressions (the sulci). These explain why brains look folded.

hippocampus Part of the brain that is extremely important for learning, memory, and navigation.

hydrophone An underwater microphone.

IUCN status Based on recognition of threatened and endangered animals (the Red List, updated annually) by a global conservation organization, the International Union for the Conservation of Nature.

kick feeding A foraging technique used by humpback whales where they pound the water surface with their tails, then dive and form a bubble cloud. This behavior is thought to be socially learned or cultural.

lateral On the side of the body or the vertical plane of a structure.

lobtailing Behavior that involves forceful slapping of the flukes against the water; also referred to as tail-slapping. Lobtailing likely also functions in non-verbal communication.

mass stranding Three or more individuals of the same species that intentionally swim or are unintentionally trapped ashore by waves or receding tides. In some cases, mass strandings of whales have been linked to military sonar.

matriline/matrilineal unit Female-centered unit with a mother, her offspring, grandoffspring and even great-grandoffspring. It is based on female kinship since the male's offspring are typically not in the group or unit.

melon Bulbous forehead of toothed whales containing fats used to focus echolocation sounds. It is the part of the head through which the clicks travel from the phonic lips into the water.

mud-ring and mud-plume feeding A behavior common among some bottlenose dolphins where they beat their tails to create a mud ring, or a plume, which forms a mud-net around prey that the dolphins can concentrate together and feed on.

mysticetes One of two major suborders of cetaceans; also called baleen whales the group includes fin and humpback whales.

neocortical convolution Folding of the neocortex (the brain's new outer layer) that increases the surface area. Associated with higher cognitive function.

odontocetes One of two major suborders of whales named for their possession of teeth; includes dolphins, porpoises, beaked whales, and sperm whales.

olfactory bulb The neural structure of the brain used in olfaction (smell). Virtually absent in cetaceans.

passive acoustic monitoring Using hydrophones researchers can record the sounds of marine mammals in an area over long periods of time.

pelagic Referring to the open ocean.

philopatry Tendency of an individual to remain with the social group or in the location of their birth for life. Usually one sex leaves their natal group when approaching maturity, but some cetaceans have bisexual philopatry where both sexes stay in the natal area or group for life.

phonic lips The structure that vibrates between the air sacs in the blowhole and the melon to produce both echolocation clicks, burst pulse sounds, and whistles that cetaceans use for communication and foraging. Also called the monkey lips.

pod Stable social group of cetaceans. Most whale pods are family groups. For example, killer whale pods comprise related females and their offspring, known as a matriline.

porpoise Family of small odontocetes that are characterized by an indistinct beak, robust body, and spade-shaped teeth.

resident Refers to an ecotype of killer whale with a distinctive morphology, genetics, behavior, and ecology, e.g. resident killer whales differ from transients in feeding exclusively on fish (especially salmon and trout). Resident killer whales live in waters along the west coast of North America. They are often referred to as "fish-eating killer whales" rather than residents since they are really no more residential to an area than transients.

rostrum see beak.

school Structured social group observed in odontocetes characterized by at least some long-term associations of individuals. Examples are various delphinid species. It usually refers to large groups.

sonar High-frequency sound system used by toothed whales during echolocation.

spectrogram A graphic representation of the frequency, intensity, variation, and duration of a sound or series of sounds.

toothed whales One of two major suborders of whales that possess teeth. See also odontocetes.

transient Refers to an ecotype of killer whale that has a large range and feeds almost exclusively on marine mammals. They range along the west coast of North America and differ from residents (fish-eating killer whales) in morphology, genetics, and ecology. They are often referred to as "mammal-eating killer whales" rather than transients.

umwelt The experience of being; how an organism experiences the world including all senses (auditory, visual, proprioceptive, olfactory, etc.) and cognition.

vocal learning A specific type of learning where the organism can modify the structure of a call during development based on social experience. Birds, humans, and cetaceans are good vocal learners. Most mammals are not.

ventral An animal's underside including chest, belly, and abdomen.

wave-washing Killer whales generate a pressure wave that "washes" their prey, seals usually, from rocks and ice floes.

RESOURCES

BOOKS

Mann, J., Connor, R. C., Tyack, P. L. & Whitehead, H., eds (2000). *Cetacean Societies: Field Studies of Dolphins and Whales*. Chicago, University of Chicago Press.

Raghanti, M. A., Munger, E. L., Wicinski, B., Butti, C., Jacobs, B. & Hof, P. R. (2017). Comparative structure of the cerebral cortex in large mammals. *Evolution of Nervous Systems*, vol. 2 *Mammals*, J. Kaas ed., Oxford, Elsevier, pp. 267–289.

Shumaker, R. W., Walkup, K. R., Beck, B. B. & Burghardt, G. M. (2011). *Animal Tool Behavior: The Use and Manufacture of Tools by Animals*. Baltimore: Johns Hopkins University Press.

PAPERS & PERIODICALS

Butti, C., Janeway, C. M., Townshend C., Wicinski, B. A., Reidenberg, J. S., Ridgway, S. H., Sherwood, C. C., Hof, P. R. & Jacobs B. (2015). The neocortex of cetartiodactyls: I. A comparative Golgi analysis of neuronal morphology in the bottlenose dolphin (*Tursiops truncatus*), the minke whale (*Balaenoptera acutorostrata*), and the humpback whale (*Megaptera novaeangliae*). *Brain Structure & Function*, 220: 3339–68.

Butti, C., Fordyce, E. R., Raghanti, M. A., Gu, X., Bonar, C. J., Wicinski, B. A., Wong, E. W., Roman, J., Brake, A., Eaves, E., Spocter, M. A., Tang, C. Y., Jacobs, B., Sherwood, C. C. & Hof, P. R. (2014). The cerebral cortex of the pygmy hippopotamus, *Hexaprotodon liberiensis* (Cetartiodactyla, Hippopotamidae): MRI, cytoarchitecture, and neuronal morphology. *Anatomical Record* 297: 670–700.

Butti C. & Hof, P. R. (2010) The insular cortex: a comparative perspective. *Brain Structure & Function*, 214: 477–493.

Butti, C., Sherwood, C. C., Hakeem, A. Y., Allman, J. M. & Hof, P. R. (2009). Total number and volume of von Economo neurons in the cerebral cortex of cetaceans. *Journal of Comparative Neurology*, 515: 243–59.

Cantor, M., Shoemaker, G. L., Cabral, R. B., Flores, C. O., Varga, M. & Whitehead, H. (2015) Multilevel animal societies can emerge from cultural transmission. *Nature Communications*, 6: 8091.

Garland, E. C., Goldizen, A. W., Rekdahl, M. L., Constantine, R., Garrigue, C., Daeschler Hauser, N., Poole, M. M., Robbins, J. & Noad, M. J. (2011). Dynamic horizontal cultural transmission of humpback whale song at the ocean basin scale. *Current Biology*, 21 (8), 688.

Geisler J. H., McGowen, M. R., Guang, Y. & Gatesy, J. (2011). A supermatrix analysis of genomic, morphological, and paleontological data from crown Cetacea. *BMC Evolutionary Biology* 11: 112.

Harley, H. E., Putman, E. A. & Roitblat, H. L. (2003). Bottlenose dolphins perceive object features through echolocation. *Nature*, 424, 667–669.

Harley, H. E. & DeLong, C. M. (2008). Echoic object representation by the bottlenose dolphin. *Comparative Cognition and Behavior Reviews*, 3, 46–65.

Hof, P. R. & Van der Gucht, E. (2007). The structure of the cerebral cortex of the humpback whale, *Megaptera novaeangliae* (Cetacea, Mysticeti, Balaenopteridae). *Anatomical Record*, 290, 1–31.

Janik, V. M. (2009). Acoustic communication in delphinids. *Advances in the Study of Behavior* 40:123–157.

Janik, V. M. & Sayigh, L. S. (2013). Communication in bottlenose dolphins: 50 years of signature whistle research. *Journal of Comparative Physiology A* 199:479–489.

Janik, V. M. (2013). Cognitive skills in bottlenose dolphin communication. *Trends in Cognitive Sciences* 17: 157–159.

McDonald, M. A., Hildebrand, J. A. & Mesnick, S. (2009). Worldwide decline in tonal frequencies of blue whale songs, *Endangered Species Research* (9) 13–21.

Mann, J. & Patterson, E. M. (2013). Tool use by aquatic animals. *Philosophical Transactions of the Royal Society of London*. Series B, *Biological Sciences*, 368 (1622), 20120424. http://doi.org/http://dx.doi.org/10.1098/rstb.2012.0424

Schwacke, L. H., Smith, C. R., Townsend, F. I., Wells, R. S., Hart, L. B., Balmer, B. C., Collier, T. K., De Guise, S., Fry, M. M., Guillette Jr., L. J., Lamb, S. V., Lane, S. M., McFee, W. E., Place, N. J., Tumlin, M. C., Ylitalo, G. M., Zolman, E. S. & Rowles, T. K. (2014). Health of common bottlenose dolphins (*Tursiops truncatus*) in Barataria Bay, Louisiana, following the *Deepwater Horizon* Oil Spill. *Environmental Science & Technology*, 48 (1), 93–103.

Williams, R. & Lusseau, D. (2006). A killer whale social network is vulnerable to targeted removals. *Biology Letters* 2(4), 497–500.

WEB SITES

Cascadia Research Collective
www.cascadiaresearch.org
Nonprofit Washington State organization that conducts research into marine mammal and bird biology, behavior, ecology, and pollution ecology.

The Discovery of Sound in the Sea
www.dosits.org
Website that provides scientific content to introduce the physical science of underwater sound and how people and animals use sound.

International Union for Conservation of Nature (IUCN) www.iucn.org
Global authority on the status of the natural world and the measures needed to safeguard it.

International Whaling Commission (IWC) iwc.int
The global intergovernmental body charged with the conservation of whales and the management of whaling. It meets annually.

National Oceanic and Atmospheric Administration (NOAA) www.nmfs.noaa.gov
Agency within the US Department of Commerce that focuses on the oceans and the atmosphere.

New Bedford Whaling Museum
www.whalingmuseum.org
Museum in Massachusetts that focuses on the history of the international whaling industry.

Marine Conservation Society (MCS)
www.mcsuk.org
Charity tasked with protecting the seas, shores, and wildlife of the UK.

Monterey Bay Aquarium
www.montereybayaquarium.org
Nonprofit public aquarium in California.

The Sea Mammal Research Unit
www.smru.st-andrews.ac.uk
Marine science research organization in Scotland that undertakes fundamental research on sea mammals.

Shark Bay Dolphin Project
www.monkemiadolphins.org
Janet Mann's long-term research study that tracks over 1,600 dolphins in Shark Bay, Western Australia, throughout their lives.

Woods Hole Oceanographic Institution
www.whoi.edu
Nonprofit research facility dedicated to the study of all aspects of marine science and engineering and to the education of marine researchers. Linked to the wesite is the Watkins Marine Mammal Sound Database: **www.whoi.edu/watkinssounds**

NOTES ON CONTRIBUTORS

CONSULTANT EDITOR

Janet Mann is professor of biology and psychology at Georgetown University, Washington, D.C., where she currently is also vice provost for research. Janet has studied dolphin behavior for more than 30 years and her extensive research into the bottlenose dolphins of Shark Bay, Western Australia has been turned into dozens of documentaries and an award-winning children's book. In 2015, she launched a new study of bottlenose dolphins in the Potomac River and Chesapeake Bay. Janet is coeditor of *Cetacean Societies: Field Studies of Dolphins and Whales*, published by the University of Chicago Press.

AUTHORS

Camilla Butti has a PhD in biomedical and comparative veterinary sciences. After working as an independent consultant she was first a research coordinator and now a visiting scientist at the Fishberg Department of Neuroscience at the Icahn School of Medicine at Mount Sinai in New York. She has published several scientific papers on comparative brain evolution and on the cortical structure and specializations of the brain of cetaceans and other large-brained mammals using both qualitative and quantitative techniques.

Heidi E. Harley has been professor of psychology at New College of Florida for more than 20 years and has been studying dolphins for more than 30. Harley has authored and co-authored numerous scientific articles on dolphin cognition and has worked with dolphins at a dozen facilities. She finds that studying the dolphin mind offers job security: the questions are many and the answers elusive.

Patrick Hof is the Regenstreif Professor of Neuroscience and vice-chair of the Fishberg Department of Neuroscience at the Icahn School of Medicine at Mount Sinai in New York. He has published over 500 scientific papers and has established an international reputation in quantitative approaches to human neuropathology, systems neuroscience, and brain evolution. He is leading a large-scale investigation of the evolutionary organization of the structure of the cerebral cortex in mammals. Patrick is also the editor-in-chief of the *Journal of Comparative Neurology* and a senior editor of several neuroscience journals.

Vincent Janik is professor of biology and director of the Scottish Oceans Institute at the University of St. Andrews in the UK. He is editor of the book series *Animal Signals and Communication* and is on the editorial board of the journal *Animal Cognition*. He has studied marine mammal communication and cognition for over 25 years, working with wild and captive marine mammals around the globe.

Eric Patterson is a behavioral ecologist with a bachelor's degree from the University of Colorado and a PhD from Georgetown University, Washington, D.C. He studied dolphin tool use in Shark Bay, Western Australia for nearly 10 years and taught courses in Marine Biology and Animal Behavior at Georgetown University. He now works for NOAA Fisheries helping to protect species listed as Endangered or Threatened under the US Endangered Species Act.

Andrew Read is the Stephen A. Toth Distinguished Professor of Marine Biology and director of the Duke University Marine Laboratory, in Beaufort, North Carolina. Andy conducts field research and works actively to conserve marine mammals throughout the world. He has served as a member of the Cetacean Specialist Group of the IUCN, the Scientific Committee of the International Whaling Commission, and the International Committee for the Recovery of the Vaquita. In March 2015 Andy was nominated by President Obama to serve as Chair of the US Marine Mammal Commission.

Luke Rendell (@_lrendell) studies the evolution of communication and learning, with interests from marine conservation, through fish, whale, and dolphin behavior, to human cultural evolution. After a PhD with Hal Whitehead in Canada he moved to the University of St. Andrews in the UK, where he is now a reader in biology affiliated with the Centre for Social Learning and Cognitive Evolution and the Sea Mammal Research Unit.

Laela Sayigh is a research specialist at the Woods Hole Oceanographic Institution and a visiting assistant professor at Hampshire College, both in Massachusetts. She has studied communication in whales and dolphins for over 30 years in species ranging from dolphins to blue whales. She has authored and co-authored numerous scientific articles and is also involved in outreach efforts for students and the general public.

Hal Whitehead is a professor at Dalhousie University, in Halifax, Nova Scotia, Canada. He uses a 40-foot (12-meter) ocean-going sailing boat to study the deep-diving whales of offshore waters—the sperm and northern bottlenose. His research focuses on the behavior, social organization, and transmission of culture among these animals, as well as on their ecology, population biology, and conservation. He also works on general methods of analyzing animal societies and cultural evolution, in theory and in practice.

Bryant Austin is an American photographer most recognized for his life-size portraits of whales. His collection of twenty-six prints are representative of over four years of fundraising and two years of actual field work with three whale species. Austin floats alone in the ocean on snorkel, observing whale behaviors from a distance. He waits until he is approached by a whale and begins photographing them when they are less than two meters from his camera. His work has exhibited internationally with shows in Norway, Japan, and Australia. His first book, entitled *Beautiful Whale*, was published in 2013. Austin was born in Sacramento, California, in 1969 and currently resides in Carmel, California.

Ewa Krzyszczyk helped enormously in finding dolphin photographs used in this book and took many of these herself. Dr. Krzyszczyk has worked on the Shark Bay Bottlenose Dolphin Project for 11 years, first as an assistant, then as a graduate student of Dr. Mann's, and now as post-doctoral scientist. She has authored or co-authored a dozen articles on dolphins and has expertise in the study of juvenile dolphins, a life history stage we still know little about.

INDEX

ACKNOWLEDGMENTS

I would like to thank Adam Pack and Robin Baird for helping us on images. I would also like to thank my many students and colleagues on the Shark Bay Dolphin Research Project who have helped make this the wonderful long-term study it is. Finally, I would like to thank my husband, Tom DeMuth, who loves and supports his workaholic wife. **Janet Mann**

PICTURE CREDITS

Every effort has been made to trace copyright holders and obtain permission. The publishers apologize for any omissions and would be pleased to make any necessary changes at subsequent printings.

Alamy Stock Photo /François Gohier/VWPics 2, /PjrStudio 38, /Frans Lanting Studio 51, /Minden Pictures 58, /Jorge Sanz/Pacific Press/Alamy Live News 64-65, /Arco Images GmbH 97t, / WaterFrame 103, /Jeff Rotman 111t, /imageBROKER 138, /Arterra Picture Library 151b, /Minden Pictures 162, /Loetscher Chlaus 167
Jenny Allen 132t, 132c, 132b
Bryant Austin 12-13, 24-5, 36-7, 66-7, 92-3, 122-3, 144-5, 160-1
Robin Baird 19br
Leighton de Barros 22r
Camilla Butti & Bridget Wicinski 32
Maurício Cantor 121
Chicago Zoological Society's Sarasota Dolphin Research Program (taken under National Marine Fisheries Service Scientific Research Permit No. 15543) 75b, 78t
Galveston Lab, National Marine Fisheries Service/Ron Wooten 178
Getty Images /Flip Nicklin/Minden Pictures 4-5, 23bl, 62, 130t, 135, 185, /Gerard Soury 6-7, /Juan Carlos Muñoz 9, /Ullstein Bild 15, 59, /Tony Korody/ Sygma 17, /Jonathan Blair 23br, /Barcroft Media 30, 100, /DEA /A. DAGLI ORTI/De Agostini 39, /Wild Horizons/UIG 41, /John Bryson 44, /Ed Kashi 46, / George F. Mobley/National Geographic 53t, /Ben Van Hook/The LIFE Images Collection 53b, /Dmitry Miroshnikov 54-5, /Kerstin Meyer 69, /Wolfgang Kaehler 70, /Paul Nicklen/National Geographic 76, /Chris Newbert/Minden Pictures 82 /Alexander Safonov 88, /Doug Perrine 89, /Paul Sutherland 95b, /Kevin Schafer/Minden Pictures 97b, /Bob Chamberlin/Los Angeles Times 99br, /MCT 104, /Mike Parry/ Minden Pictures 105b, /Panoramic Images 117, / by wildestanimal 120, 139, /Timothy Allen 127b, /Auscape 130b, /Peter Chadwick 137, / Cyril Ruoso/Minden Pictures 146, /Kevin Schafer/ Minden Pictures 149b, /Brian J. Skerry 152, /Luciana Whitaker 168, /Richard Smith/Sygma 169, /Anthony Wallace/AFP 172b, /UniversalImagesGroup 176, / Tony Ashby/AFP 183b
Quincy Gibson 140
Jonathan H. Giesler 11
Heidi E. Harley 50
iStock /Andrew_Howe 16, /Yann-HUBERT 44-5 (background), /flyingrussian 56-7 (background), /Getty Images Plus 60-1 (background), /duncan1890 166
Vincent Janik 87
Ewa Krzyszczyk 19bl, 22l, 108b, 111b, 113t, 115, 119b, 141, 142r, 153l, 153r, 154l, 157t, 157b, 158
Claudia Kuster 84t
Louisiana Department of Wildlife and Fisheries/ Mandy Tumlin 179b
Madison Miketa 108t
Janet Mann 19tr, 20-1, 95t, 113b, 116, 142l, 143
Tony Martin/Projeto Boto 148
Nature Picture Library/Tony Wu 81, /Roland Seitre 150, /Doc White 184
Silvana Neves 119t
NHPA/Avalon 107t
NOAA /Kevin Lino NOAA/NMFS/PIFSC 63, /Christin Khan/NEFSC 107b, /NOAA Fisheries/ Robert Pitman 152r, /Paula Olson 165, /NOAA Fisheries/Barbara Taylor 175b
Image supplied with kind permission by **NOAA/C. Faesi, Proyecto Vaquita, copyright Omar Vidal** 172t
Susan Parks 73bc, 73b
Eric Patterson 127t, 154b, 155 (inset images 1, 2, 3)
Douglas Peebles 43
Heinz Plenge 149t
Royal Society (from Williams, R. & Lusseau, D., A killer whale social network is vulnerable to targeted removals, *Biology Letters* (2006) 2(4): 497-500. By permission of the Royal Society.) 99l
Luke Rendell 133l, 134l
© www.savethewhales.org 175t
Laela Sayigh & Vincent Janik spectrograms 77, 79, 83, 85
Barry Shoesmith 154-5 (background image)
Shutterstock /Uwe Bergwitz 19tl, /Aleksandr Kutakh 20-21, 140-43, 175 (background), /Willyam Bradberry 71b, /Joost van Uffelen 77b, /wildestanimal 82r, / Peter Asprey 83tl, / Monika Wieland Shields 99tr, / Durk Talsma 105t, /Everett Historical 124, /Seb c'est bien 125, /Rex 164 /Rich Carey endpapers
Scott Tuason 154t, 155 (inset image 4)
Danielle Waples 181b
Watkins Marine Mammal Sound Database /Woods Hole Oceanographic Institution 73t, 73tc, 75t
Image supplied with kind permission by **Mark Xitco** (Source: *International Journal of Comparative Psychology*, 26: 119, Kuczaj II, Stan A. (2013)) 49